Compendium on MICRONUTRIENT FERTILISERS IN INDIA

Crop Response & Impact, Recent Advances and Industry Trends

2nd Edition

Edited by
Dr. Shama Zaidi & Dr. Rahul Mirchandani

2022

PARTRIDGE

Copyright © 2022 by Dr. Shama Zaidi & Dr. Rahul Mirchandani.

ISBN: Softcover   978-1-5437-0864-6
          eBook   978-1-5437-0863-9

All rights reserved. No part of this book may be used or reproduced by any means, graphic, electronic, or mechanical, including photocopying, recording, taping or by any information storage retrieval system without the written permission of the author except in the case of brief quotations embodied in critical articles and reviews.

Because of the dynamic nature of the Internet, any web addresses or links contained in this book may have changed since publication and may no longer be valid. The views expressed in this work are solely those of the author and do not necessarily reflect the views of the publisher, and the publisher hereby disclaims any responsibility for them.

Print information available on the last page.

To order additional copies of this book, contact
Partridge India
000 800 919 0634 (Call Free)
+91 000 80091 90634 (Outside India)
orders.india@partridgepublishing.com

www.partridgepublishing.com/india

Preface

The importance of balanced crop nutrition as a national imperative, essential to double farm income, is well recognized over the decades. With the efforts of the Micro Fertilizers and Specialty Plant Nutrition industry, millions of Indian farmers have been made aware of the good agricultural practices that lead to sustainably feeding their crops with all essential plant nutrients. The role of each micronutrient has been carefully demonstrated using demos, trials and extensive field research.

To chronicle the industry wide current body of research on micronutrient fertilisers, their importance, availability, methods of supplementation and role in integrated nutrient management, IMMA presents with great pleasure this Compendium on various aspects of Micronutrient Nutrition.

The Indian Micro-Fertilizers Manufacturers Association (IMMA) was established in 1983 to represent the interests of the micronutrient manufacturers throughout the country. Our members have been working with the farming communities for almost 40 years and have played a vital role in making balanced plant nutrition a National Imperative. IMMA has three major focus areas of work: Policy Advocacy with Government, Knowledge and Skills Training of staff and stakeholders across India, as well as developing unity and Fellowship amongst members.

I am delighted that the second edition of this Compendium has been carefully curated and will serve as a comprehensive and authoritative pool of information on various aspects of Micronutrient fertilisers, covering a wide range of topics from the basics of detecting and correcting deficiencies, salient features of different variants of micronutrient fertilisers, new developments like Nano fertilisers, real time nutrient assessment, use of drones, customised and innovative fertilisers. This edition further documents the impact of micronutrients on human health and also lists state specific grades and presents industry pooled data on usage of Micronutrients at various University trials.

With this rich pool of latest information, farmers, industry colleagues and the scientific community will undoubtedly find this Compendium as an essential reference for everything related to the Micro Fertiliser Industry. The agricultural demonstrators, policy makers and agri extension workers would also find this Compendium very useful.

The effort to gather all the data and analysis presented in the Compendium has been tremendous and IMMA gratefully acknowledges the time and energy put in by Dr. Shama Zaidi and her colleagues Mr. Jayapradeep Subramanian, Ms. Anagha Gaikwad, Ms. Karishma Talekar, Mr Amey Rane and Mr. Sachin Sawant at Aries Agro Ltd. This Compendium is available on IMMA's website in e-book form and also in hard copy.

IMMA shall be open to receive further information and research data from its members and other industry colleagues to update this Compendium for future editions. The usage of this volume and its widespread circulation will be the best reward that our Association shall receive for this massive effort.

Sincerely
Dr. Rahul Mirchandani
President
Indian Micro-Fertilisers
Manufacturers Association

6th May, 2022
New Delhi

Acknowledgements

Grateful thanks are due to

M/s ARIES AGRO LIMITED for permitting **Dr. Shama Zaidi** to complete this task.

All authors and publishers pertaining to the articles from where several information are taken and **All IMMA Members and Agro Inputs Manufacturers Association (India)** who have forwarded their research information for this compendium.

Indian Micro-Fertilizers Manufacturers Association

Bringing all round improvement, friendship and unity amongst the manufacturers of micronutrients across India, since over 35 years

About IMMA

The Indian Micro-Fertilizers Manufacturers Association (IMMA) works for the welfare of manufacturers engaged in manufacture of micro nutrients. IMMA has a PAN India presence and is has been a united voice for Indian micronutrient manufacturers for more than 35 years. IMMA has a seat on key statutory policy making bodies, like the State Fertilizer Committees and the Bureau of Indian Standards. IMMA is also a member of Fertilizer Association of India and the Confederation of Indian Industry.

IMMA's flagship National Crop Nutrition Summits

IMMA brings together all stakeholders - Government, Scientists, Research Institutions, Farmers, other Industry Associations for our flagship National Crop Nutrition Summits for a comprehensive dialogue with IMMA members and based on the deliberations, a Communiqué is released summarizing specific challenges and proposed solutions.

IMMA strives to share best practices amongst members, by bringing together top scientists and researchers for expert technical sessions, visits to best-in-class labs to standardize analytical methods and publishing latest research papers in the periodical IMMA Bulletins.

Policy Advocacy

IMMA continuously interacts with concerned Ministers, Agricultural Commissioner & policy makers at the Central and State Governments.

AIMS AND OBJECTIVES OF IMMA

POLICY ADVOCACY

1. To arrange through the Association, healthy dialogue between various institutions as and when necessary.
2. To make efforts for the welfare of related institutions, farmers, dealers, etc. and to unite them to safeguard the interest of common nature so as to bring all round prosperity.
3. To bring about the understanding between government, manufacturers and consumers for the benefit of all.

KNOWLEDGE SHARING

4. To hold agricultural seminars, Exhibitions, Demonstrations, etc.
5. To improve technical knowledge of the members.
6. To educate, guide and advice on various Acts like Fertilizer Control Order, Weight & Measures Act and Essential Commodities Act, etc.
7. To create consciousness amongst the public and to disseminate and help in dissemination of latest crop nutrition techniques.
8. To keep the Members informed from time to time about Acts, Rules, Policies, Experiments, Research, Plans, etc.

FELLOWSHIP

9. To bring all round improvement, friendship and unity among the manufacturers of micronutrients
10. To help to solve the disputes, difference of opinion etc. between the members of the Association.
11. To undertake any other activity necessary in the interest of the Association as maybe deemed necessary.

Contents

LIST OF ANNEXURES XV

CHAPTER 1: MICRONUTRIENTS – IMPORTANCE, AVAILABILITY AND UPTAKE 1

Section 1.1 Micronutrients - General Introduction 1
Section 1.2 Micronutrients - Roles And Deficiency Symptoms 10
Section 1.3 Detecting Micronutrient Deficiencies: Soil Analysis And Tissue Analysis 23
Section 1.4 Micronutrient Status Of Indian Soils 26
Section 1.5 Micronutrient Management For Higher Crop Production 35

CHAPTER 2: MICRONUTRIENT FERTILISERS 39

Section 2.1 Fertiliser Control Order 39
Section 2.2 State Specific Grades 42
Section 2.3 Economics Of Fertiliser Use 50
Section 2.4 Impact Of Fertiliser Application On Crop Yields 55

CHAPTER 3: METHODS OF SUPPLEMENTATION 61

Section 3.1 Specialty Fertilisers 61
    Subsection 3.1 a Water Soluble Fertilisers 63
    Subsection 3.1 b Slow And Controlled Release Fertilisers 65
    Subsection 3.1 c Fortified Fertilisers 68
    Subsection 3.1 d Liquid Fertilisers 69
    Subsection 3.1 e Stabilized Fertilisers 70
    Subsection 3.1 f Customized Fertilisers 70

Article 1. Need for Specialty Nutrition 72
    Section 3.2 Chelation And Chelated Nutrients 77

Article 2. Balanced Crop Nutrition Using Metal Chelates 81
    Section 3.3 Nano-Fertilisers 88

Article 3. Nanotechnology in The World Of Agriculture 99

CHAPTER 4: INTEGRATED NUTRIENT MANAGEMENT...105

    Section 4.1 IPNM- Introduction 105
    Section 4.2 4R Nutrient Stewardship For Fertilizer Management...... 109

Article 4. Integrated Nutrition Management for Alleviation of Abiotic Stress In Crops 116
    Section 4.3 Precision Farming And Site-Specific Nutrient Management 122
    Section 4.4 Role Of Agricultural Extension 127

CHAPTER 5: MICRONUTRIENTS- IMPACT ON HUMAN HEALTH 133

    Section 5.1 Micronutrient Deficiencies In Soils, Plants, Animals And Humans 133

Article 5. Micronutrients: Striding Towards Food and Nutrition Security 140
    Section 5.2 Micronutrient-Related Malnutrition 144
    Section 5.3 Impact Of Micronutrient Deficiencies And Toxicities On Crop Productivity And Animal And Human Health 151
    Section 5.4 Strategies To Improve Micronutrient Nutrition In Animals And Humans 162
    Section 5.5 Biofortification- Producing Micronutrient-Rich Food 166
    Section 5.6 Field Crops As A Rich Source Of Minerals 173

Article 6. Impact of fertilizers with special reference to micronutrients on environment pertaining to human and animal health 178

CHAPTER 6: RECENT ADVANCES IN AGRICULURE TECHNOLOGY 185

Section 6.1 Recent Advances In Agriculture.. 185
Section 6.2 Real-Time Assessment Of Nutrient Status 190
    Subsection 6.2 a Field Sensing 190
    Subsection 6.2 b Remote Sensing.. 198
Section 6.3 Application Of Drone Technology In Agriculture 207

Article 7. AGRICULTURE DRONE- Future of farm activities............. 211

CHAPTER 7: AGRICULTURAL MICRONUTRIENTS- INDUSTRY TRENDS... 215

Section 7.1 Agricultural Micronutrients- Market In India 215

BIBLIOGRAPHY.. 223

List Of Annexures

No	Chelate	University	Data/Report provided by	Page No
1	Manganese EDTA	National Horticultural Research and Development Foundation, Nasik	Aries Agro Limited	251
2	Manganese EDTA	Mahatma Phule Krishi Vidyapeeth, Rahuri	Agro Inputs Manufacturers Association	257
3	Manganese EDTA	Dr. Balasaheb Sawant Kokan Krishi Vidyapeeth, Dapoli	Agro Inputs Manufacturers Association	260
4	Manganese EDTA	Vasantrao Naik Marathwada Krishi Vidyapeeth, Parbhani	Agro Inputs Manufacturers Association	262
5	Calcium EDTA	Acharya Ranga Agricultural University, Hyderabad	Aries Agro-vet Industries	265
6	Calcium EDTA	Dr. Panjabrao Deshmukh Krishi Vidyapeeth, Akola	Agro Inputs Manufacturers Association	268
7	Calcium EDTA	Vasantrao Naik Marathwada Krishi Vidyapeeth, Parbhani	Agro Inputs Manufacturers Association	271
8	Calcium EDTA	Mahatma Phule Krishi Vidyapeeth, Rahuri	Agro Inputs Manufacturers Association	275
9	Copper EDTA	Vasantrao Naik Marathwada Krishi Vidyapeeth, Parbhani	Agro Inputs Manufacturers Association	279
10	Copper EDTA	Dr. Panjabrao Deshmukh Krishi Vidyapeeth, Akola	Agro Inputs Manufacturers Association	282

11	Magnesium EDTA	Dr. Panjabrao Deshmukh Krishi Vidyapeeth, Akola	Agro Inputs Manufacturers Association	289
12	Magnesium EDTA	Mahatma Phule Krishi Vidyapeeth, Rahuri	Agro Inputs Manufacturers Association	291
13	Magnesium EDTA	Dr. Balasaheb Sawant Kokan Krishi Vidyapeeth, Dapoli	Agro Inputs Manufacturers Association	293
14	Ferrous EDTA	Mahatma Phule Krishi Vidyapeeth, Rahuri	Agro Inputs Manufacturers Association	298
15	Ferrous EDTA	Vasantrao Naik Marathwada Krishi Vidyapeeth, Parbhani	Agro Inputs Manufacturers Association	302
16	Multi micronutrient formulation	National Horticultural Research and Development Foundation, Nasik	Aries Agro Limited	304
17	Multi micronutrient formulation	National Horticultural Research and Development Foundation, Nasik	Aries Agro Limited	308
18	Liquid chelated micronutrient mixture	National Horticultural Research and Development Foundation, Nasik	Aries Agro Limited	313
19	Amino acid chelated Zn and Zinc-EDTA comparative	National Horticultural Research and Development Foundation, Nasik	Aries Agro Limited	317
20	Amino acid chelated micronutrient mixture	National Horticultural Research and Development Foundation, Nasik	Aries Agro Limited	322
21	Yield Enhancers	ICAR-IIHR	Karnataka Agro Chemicals	326
22	Comprehensive Trial Report (1975-1998)		Aries Agro Limited	334

List of Tables

1.1 Essential mineral nutrients for plants 3
1.2 Micronutrient cations and anions 4
1.3 Nutrient uptake by various crops 5
1.4 Functions of various nutrients in plants 10
1.5 Nutrient Deficiency Symptoms in common Crop groups 15
1.6 Nutrient mobility in Soil and Plants 22
2.1 Recommended grade as per Government of Andhra Pradesh 42
2.2 Recommended grade as per Government of Assam 42
2.3 Recommended grade as per Government of Bihar 43
2.4 Recommended grade as per Government of Chhattisgarh
    Notification No 4919 44
2.5 Recommended grade as per Government of Chhattisgarh
    Notification No 4022 44
2.6 Recommended grade as per Government of Gujarat 44
2.7 Recommended grade as per Government of Himachal Pradesh 45
2.8 Recommended grade as per Government of Jharkhand 45
2.9 Recommended grade as per Government of Karnataka 46
2.10 Recommended grade as per Government of Madhya Pradesh 46
2.11 Recommended grade as per Government of Maharashtra 46
2.12 Recommended grade as per Government of Orissa 47
2.13 Recommended grade as per Government of Tamil Nadu 48
2.14 Recommended grade as per Government of Rajasthan 48
2.15 Recommended grade as per Government of Telangana 48
2.16 Recommended grade as per Government of Uttaranchal 49
2.17 Recommended grade as per Government of Uttar Pradesh 49
2.18 Recommended grade as per Government of West Bengal 49
2.19 Cost- Benefit analysis of chelated micronutrients 53
2.20 Average crop responses to Zinc in field experiments 56
2.21 Average crop responses to Iron in field experiments 57
2.22 Summary of Crop responses to Boron application 58

3.1 Forms of N in NP / NPK complexes 64
3.2 Customised fertiliser formulations available in India 71
3.3 Impact of nanofertilisers on productivity of different crops under
    varying pedo-climatic conditions 94
3.4 Characteristics imparted by nanomaterials in different crops 95
4.1 Components of the 4R Nutrient Stewardship system 110
5.1. Critical concentration of micronutrients in crop plants 136
5.2 Prevalence of the three major micronutrient deficiencies by WHO region ... 147
5.3 Selected micronutrient deficiencies, their risk factors and effects 148
5.4 Ferti-fortification of minerals in various crops 169
5.5. Foliar application of micronutrients for mineral fertilization 170
5.6 Seed treatment with various micronutrients 171

List of Figures

1.1: Effect of soil pH on availability of plant nutrients 8
1.2: Some common deficiency symptoms 13
1.4: Portion of the plant where nutrient deficiency symptoms are first observed 21
1.5: Relationship between plant growth and health and amount of nutrient available. 23
1.6: Micronutrient deficiency in Indian soils 27
1.7: Zinc deficiency status in soils of India 29
1.8: Iron deficiency status in soils of India 30
1.9: Manganese deficiency status in soils of India 31
1.10: Copper deficiency status in soils of India 32
1.11: Boron deficiency status in soils of India 33
1.12: Multimicronutrient and Secondary nutrient deficiency in soil 34
3.1a: Slow and Controlled Release fertiliser 65
3.1b: Slow and Controlled Release fertiliser 68
3.2: Role of chelates in nutrient availability 78
3.3: Potential smart fertiliser effects in the soil-plant system 90
3.4: Potential entry points of nanoparticles into plants 91
4.1: Performance applying farmyard manure (FYM) and fertilisers as integrated plant nutrient management vis-à-vis fertiliser alone in rice 106
4.2: Economic, social and environmental goals related to nutrient management 111
4.3 Performance indicators 111
4.4: Impact of different fertilizer placement practices for movement of nutrients into the soil 114
4.5: Various component practices and technologies commonly associated with specific precision agriculture systems 126
5.1: Visual micronutrient deficiency symptoms of various crops 137
5.2: Effects of micronutrient malnutrition on human health. 146
5.3: Different approaches to achieve biofortification 167
5.4: Different modes of mineral fertilization 168

5.5: *Different amendments under agronomic management practices to enhance mineral content in plant* 172

6.1: *Common important nutrient deficiencies in plants* 192

6.2: *Nitrogen leaf color chart (LCC)* 193

6.3a: *Hand held sensor* 194

6.3b: *Machine mounted canopy sensor* 194

6.4: *The SPAD chlorophyll meter* 195

6.5a: *The Green Seeker system in use in the field* 196

6.5b: *Operation of the GreenSeeker on-the-go sensing system* 196

6.6: *Field Scout Green Index + Nitrogen App and Board* 197

6.7: *Typical Spectral Reflectance curves for vegetation, dry bare soil and water* 198

6.8 i: *Aerial Photo* 200

6.8 i, ii and iii. *Processed images based upon aerial photography showing the N deficient areas identified in the early season aerial photograph accurately predicted yield losses* 201

6.8 ii: *Yield loss map based on yield monitor data* 201

6.8 iii: *Yield loss map predicted from the aerial photo* 201

6.9: *Use of drones for spraying nutrients and pesticides* 208

6.10: *Use of drones for field imaging* 209

7.1: *Agricultural Micronutrients Market in India- Market Landscape* 216

7.2: *Key Growth Drivers- Overview* 217

7.3: *Agricultural Micronutrient Market – Based On Type* 218

7.4: *Agricultural Micronutrient Market – By Crop Type Analysis* 220

Micronutrients – Importance, Availability And Uptake

1.1 MICRONUTRIENTS - GENERAL INTRODUCTION

All plants require nutrients for proper growth and development in sufficient quantities. 16 essential nutrients have been identified and grouped according to the relative amount of each that plants need,

- *Primary nutrients*, also known as macronutrients, are those usually required in the largest amounts. They are carbon, hydrogen, oxygen. nitrogen, phosphorus, and potassium. The big three, nitrogen, phosphorus and potassium (NPK) together comprise over 75% of the mineral nutrients found in the plant.
- *Secondary nutrients* are those usually needed in moderate amounts compared to the primary essential nutrients. They are calcium, magnesium, and sulfur.
- *Micro- or trace nutrients* are required in tiny amounts compared to primary or secondary nutrients. Micronutrients are boron, chlorine, copper, iron, manganese, molybdenum, and zinc.
- Very few plants need five other nutrients: cobalt, nickel, silicon, sodium, and vanadium.

Plants take up these nutrients from the air, the soil, and the water.

ESSENTIAL V/S BENEFICIAL NUTRIENTS

Plant nutrition involves interrelationships between complex organic/inorganic elements in soil or soilless media and their role in plant physiology. It also takes into account the complex balance between essential and beneficial elements for optimum plant growth.

The term essential mineral element (or mineral nutrient) was proposed by Arnon and Stout (1939). They concluded three criteria must be met for an element to be considered essential. These criteria are:

- A plant must be unable to complete its life cycle in the absence of the mineral element.
- The function of the element must not be replaceable by another mineral element.
- The element must be directly involved in plant metabolism.

The beneficial elements have not been deemed essential for all plants but may be essential for some. They may compensate for toxic effects of other elements or may replace mineral nutrients in some other less specific function such as the maintenance of osmotic pressure. The distinction between beneficial and essential is often difficult in the case of some trace elements. The omission of beneficial nutrients in commercial production could mean that plants are not being grown to their optimum genetic potential but are merely produced at a subsistence level. A more holistic approach to plant nutrition would not only include only nutrients essential to survival but would also include beneficial mineral elements at levels necessary for optimum growth. Table 1.1 gives a list of essential mineral nutrients needed for plant growth.

Table 1.1 Essential mineral nutrients for plants

Nutrient Element	Symbol	Forms Absorbed	Typical concentration in plant dry matter
Macronutrients (Primary and Secondary)			
Nitrogen	N	NH_4^+, NO_3^-	1.5%
Phosphorus	P	$H_2PO_4^-, HPO_4^{-2}$	0.1 – 0.4%
Potassium	K	K^+	1 – 5 %
Sulphur	S	SO_4^{-2}	0.1 – 0.4%
Calcium	Ca	Ca^{+2}	0.2 – 1.0%
Magnesium	Mg	Mg^{+2}	0.1 – 0.4%
Micronutrients (Trace Elements)			
Boron	B	$H_3BO_3, H_2BO_3^-$	6 – 60 ppm
Iron	Fe	Fe^{+2}	50 – 250 ppm
Manganese	Mn	Mn^{+2}	20 – 500 ppm
Copper	Cu	Cu^+, Cu^{+2}	5 – 20 ppm
Zinc	Zn	Zn^{+2}	21 – 150 ppm
Molybdenum	Mo	MoO_4^{-2}	Below 1 ppm

NUTRIENT AVAILABILITY AND UPTAKE

Plants absorb most of the nutrients through its fine root hairs in the soil except for Carbon which is taken in through the leaf pores or the stomata.

To be able to be used up by the plant, the nutrient must be broken down to its basic form i.e. either as a positively charged ion (cation) or a negatively charged ion (anion). Complex organic compounds like dead leaves, twigs, branches, composts, manures, etc. cannot be used up unless they are broken down to their basic forms.

Mentioned in the table 1.2 below are some micronutrient cations and anions.

Table 1.2 Micronutrient cations and anions

CATIONS	
Copper	• Positively charged; binds to soil particles
Iron	• Solubility is greatest under acid conditions
Manganese	• Most likely deficient on calcareous soils or soils extremely high in organic matter where strong chelation decreases availability
Zinc	
ANIONS	
Boron	• Negatively charged; subject to leaching
Molybdenum	• In short supply in areas where they are readily leached and are not replenished by organic matter decomposition
Chlorine	

It is important to note that crops can take up nutrients only in their inorganic form though they do respond to all nutrient sources. Organic nutrients also have to be mineralized before the plants can take it up. The amount of nutrients provided by the different sources varies greatly between and within agro-ecosystems. Sustainable crop nutrition identifies and utilizes all available sources of plant nutrients.

Nutrient uptake refers to the total amount of nutrients taken up (absorbed) by a crop during its entire life cycle. The nutrient uptake pattern differs for different crops, nutrients, seasons and growth conditions. Nutrient uptake is slow in the beginning and during the ripening phase before harvest, however, the uptake is fast and rapid during the active growth and development phase. If during periods of very high / rapid nutrient uptake, the rate of nutrient supply to the roots falls behind the demand, the result is an underfed crop giving low yields.

Nutrients taken up can vary from less than 50 to 1000 kg/ha depending upon the crop, variety, nutrient and its availability, growth conditions and yield produced. Nitrogen and Potassium are absorbed in largest amounts while Molybdenum uptake is least. A general idea of the amounts of nutrients absorbed by several crops under field conditions is given in Table 1.3

Table 1.3 Nutrient uptake by various crops

Group	Crop	Main Produce	Total Uptake /t of main produce (kg)						Total Uptake /t of main produce (gms)					
			N	P_2O_5	K_2O	S	Ca	Mg	Zn	Fe	Mn	Cu	B	Mo
Cereals	Rice	Paddy	20	11	30	3	7	3	40	153	675	18	15	2
	Wheat	Grain	25	9	33	4.7	5.3	4.7	56	624	70	24	48	2
	Maize	Grain	29.9	13.5	32.8				130	1200	320	130		
	Sorghum	Grain	16.4	7.7	25.5				72	720	54	6	54	2
	Pearl Millet	Grain	31.8	17.4	61.3				40	170	20	8		
Pulses	Chickpea	Grain	60.7	9.2	39.2	8.7	18.7	7.3	57	1302	105	17		
	Pigeon pea	Grain	70.8	15.3	16	7.5	19.2	12.5	38	1440	128	31		
Oilseeds	Groundnut	Seed	58.1	19.6	30.1	7.9	20.5	13.3	208	4340	176	68		
	Mustard	Seed	32.8	16.4	41.8	17.3	42	8.7	150	1684	143	25		
	Soybean	Seed	70.7	30.9	57.7	6.7	14	7.6	192	866	208	74		
	Sesame	Seed	51.7	22.9	64	11.7	37.5	15.8	202	952	138	140		
	Sunflower	Seed	63.3	19.1	126	11.7	68.3	26.7	28	645	109	23		
	Linseed	Seed	60	18.6	54	5.6	31.2	13.1	73	1062	283	48		
Tubers	Potato	Tuber	3.3	0.9	6.2	0.4	1	1.8	9	160	12	12	50	<1
	Cassava	Tuber	5	2.3	6.8	0.4	2.7	1	45	120	45	5	15	
Fibres	Jute	Dry fibre	35.2	20.3	63.2		39.7	8	139	368	119	18		
Plantation	Tea	Leaves	178.3	3.5	115.1	10	41.7	11.5	276	2007	1933	632	101	
	Coffee	Beans	129	27	174	5			35	83	62	82		

Data Source: Several Indian published sources, Fertilizer Management by Dr HLS Tandon

FACTORS AFFECTING AVAILABILITY OF MICRONUTRIENTS

Soil reaction: -

Soil pH influences solubility, concentration in soil solution, ionic form and mobility of micronutrient in soil and consequently acquisition of these elements by plants (Fageria *et.al.*, 1997). As a rule, the availability of B, Cu, Fe, Mn and Zn usually decreases and Mo increases as soil pH increases. These nutrients are usually adsorbed onto sesquioxide soil surfaces.

Effect of clay minerals: -

The trace elements released during the course of weathering may be locked up in the crystal lattices of the clay minerals and thus become relatively unavailable. Clay minerals bind trace elements with varying degree of forces. Different binding forces account for variation in the extraction rate of different trace element cations by different reagents. Trace elements are readily adsorbed by clay minerals but displaced with difficulty. A few trace elements like B and Mo when displaced by weathering appear as complex anions. Calcium or organic complex may bind these anions (Sharma and Kumar, 2016).

Organic matter: -

Organic matter affects the availability of micronutrients. Organic materials such as manure may supply chelating agents that add in maintaining the solubility of micronutrients. Organic matter content increases the boron availability by preventing its leaching loss and bringing about its accumulation in surface soil. Addition of organic matter to well drained soils have produced varying effects on iron availability. Manganese availability is low in basic soils, high in organic matter because of the formation of unavailable chelated Mn^{2+} compounds. Action of organic matter on zinc availability is variable. When immobilization and complexation reactions of organic matter prevail, availability of soil zinc will be adversely affected. On the other hand, formation of soluble chelated compounds of zinc will enhance availability by shielding the retained zinc from fixation reactions (Sharma and Kumar, 2016).

Inter-relationship with other nutrients: -

Availability of micronutrients in soil are affected by interaction among micronutrients as well as with other nutrient elements. Metal cations including Cu^{2+}, Fe^{2+} and Mn^{2+} inhibit plant uptake of Zn^{2+} possibly because of competition for the same carrier site. High phosphorus availability induces zinc deficiency. Increased rates of potassium accentuate boron toxicity at high levels of boron. Application of acidic nitrogenous fertilizer can aggravate copper deficiencies may be due to increased aluminium levels in soil solution. Excess of nutrients such as Co, Cu, Mn, Mo, Zn and P encourage iron deficiency. Absorption and translocation of molybdenum was enhanced by application of phosphorus due to release of adsorbed MoO_4, thus making it more available to plants (Sharma and Kumar, 2016).

EFFECT OF SOIL pH ON NUTRIENT UPTAKE

Soil pH is one of the most critical factors affecting nutrient absorption and utilization. Adjusting soil pH to a recommended value can increase the availability of important nutrients. The ideal pH range will vary; however, most plants favor a mildly acidic growing environment. Plants usually grow well at pH values above 5.5. Soil pH of 6.5 is usually considered optimum for nutrient availability. Extreme pH values decrease the availability of most nutrients.

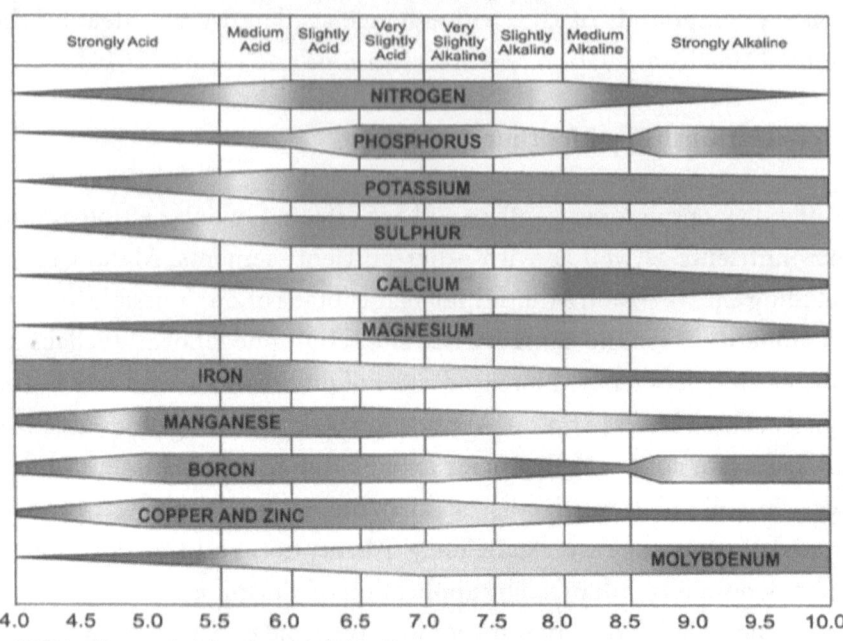

Fig 1.1: Effect of soil pH on availability of plant nutrients

Low pH reduces the availability of the macro- and secondary nutrients, while high pH reduces the availability of most micronutrients. Microbial activity may also be reduced or changed. The chart below indicates how the pH of a solution determines whether the element is more and less available to the plants

NUTRIENT REMOVAL BY CROPS

Nutrient removal is the quantity of nutrients removed in plant material harvested from the field. The extent of nutrient removal from the field depends on the plant parts which are harvested, their composition and their share in the total dry matter (biomass) production and hence cannot be equated to nutrient uptake in most cases. Crops like jute, cassava which shed their leaves before harvest return a significant part of the absorbed nutrients through leaf fall. Same is not the case with crops like leafy vegetables, root crops, etc. wherein the total biomass is harvested or when the whole plant (grain + straw) is taken away and only stubbles and roots are left behind. In such cases, nutrient removal is more or less equal to nutrient uptake.

POTENTIAL FOR ADDING NEW ELEMENTS TO THE LIST OF ESSENTIAL MINERALS

During the past 30 years, animal scientists have added many new trace elements (i.e., elements that occur in the organism in 100 pg/g amounts or less) to their list of essential elements, including As, B, Br, Cd, Cr, Pb, Li, Mo, Se, Si, Sn, V, and Ni (Van Campen, 1991). However, plant scientists have added only one new micronutrient (i.e., Ni) (Eskew, *et al.,* 1983, 1984) to their list of essential elements since 1954 when chlorine was added (Asher, 1991).

Future research should discover new essential elements for higher plants because modem sensitive analytical instrumentation (e.g., inductively coupled, Ar-plasma, mass spectrometry; laser-probe mass spectrometry; Zeeman effect, carbon furnace, atomic absorption spectrophotometry) and new plant culture techniques (e.g., chelate-buffered nutrient solutions (Taylor, *et al.,* 1990; Bell, *et. al.,* 1991; Norvell, 1991;) have been developed that make it possible for such research to be carried out efficiently at ultra-low trace metal concentrations (Norvell, 1991; Welch, 1990).

Asher (1991) reviewed the literature concerning the possible essentiality of several beneficial trace elements that may be added to the list of essential elements for higher plants if future research demonstrates that they fit the required criteria of essentiality established for higher plants Elements discussed by Asher include Na, Si, Al, Co, La, and Ce. Other trace elements reported to be essential for some organisms and deserving of further research on their possible essentiality for higher plants encompass As, Cr, Se, Li, Rb, V, Sn, Cd, Pb, and the halides I, Br, and F (Nielsen, 1990, 1992; Van Campen, 1991).

Interest in the essentiality of rare earth elements (the lanthanide series of elements from atomic number 57 to 71 [e.g., La, Ce, Pr, Nd, Pm, Sm, etc.]) has increased with reports from China indicating increased crop yields and improved product quality from applications of a fertilizer - "Nongle" - containing a mixture of La and Ce nitrates (Asher *et.al.*, 1990). These claims, however, await conformation by publication of carefully documented research reporting rigorous experimentation obtained under controlled laboratory conditions before the benefits of rare earths on plant growth or their essentiality for higher plants can be critically evaluated (Asher, 1991).

1.2 MICRONUTRIENTS - ROLES AND DEFICIENCY SYMPTOMS

Micronutrients differ in the form they are absorbed by the plant, their functions and mobility in the plant, and their characteristic deficiency or toxicity symptoms. Table 1.4 gives a summary of the various functions of plant nutrients.

Table 1.4 Functions of various nutrients in plants

Element	Function in Plant
Nitrogen	Promotes growth of leaf and stem, imparts green color, necessary to develop cell proteins and chlorophyll
Phosphorus	Early formation and growth of roots, stimulates flowering and seed development
Potassium	Formation of carbohydrates and proteins, formation and transfer of starches, sugars and oils, increases disease resistance, vigor and hardiness
Calcium	Important part of cell wall, influences intake of other plant nutrients, improves plant vigor
Magnesium	Chlorophyll formation, influences intake of other essential nutrients, translocation of phosphorus and fats
Sulphur	Promotes root growth and vigorous vegetative growth, protein formation
Iron	Essential for chlorophyll synthesis, photosynthesis, component of enzymes
Copper	Component of enzymes, helps in photosynthesis
Zinc	Component of many enzymes, essential for plant hormone balance, auxin activity, reproduction

Boron	Translocation of sugars, cell division, amino acid production
Manganese	Co-factor in many plant reactions, activates enzymes, nitrogen transformation
Molybdenum	Involved in nitrogen metabolism, essential in nitrogen fixation by legumes

COMMON SYMPTOMS FOR DIFFERENT NUTRIENT DEFICIENCIES

The first step in treating nutrient deficiency is to diagnose and describe the symptoms. Each deficiency symptom is related to some function of the nutrient in the plant. Symptoms caused by nutrient deficiencies are generally grouped into five categories:

1. **Stunted growth**

    Stunting is a common symptom for many deficient nutrients due to their varied roles in the plant. When nutrients involved in plant functions such as stem elongation, photosynthesis, and protein production are deficient, plant growth is typically slow and plants are small in stature

2. **Chlorosis**

    Chlorosis is seen in plants deficient in nutrients needed for photosynthesis and/or chlorophyll production. Chlorosis can result in either the entire plant or leaf turning light green to yellow, or appear more localized as white or yellow spotting

3. **Interveinal chlorosis**

    Like chlorosis, interveinal chlorosis is also seen in plants deficient in nutrients needed for photosynthesis and/or chlorophyll production. Interveinal chlorosis occurs when certain nutrients [B, Fe, magnesium (Mg), Mn, nickel (Ni) and Zn] are deficient.

4. **Purplish-red coloring**

   Purplish-red discolorations in plant stems and leaves are due to above normal levels of anthocyanin (a purple coloured pigment) that can accumulate when plant functions are disrupted or stressed. This symptom can be particularly difficult to diagnose because cool temperatures, disease, drought and even maturation of some plants can also cause anthocyanin to accumulate (Bennett, 1993). Certain plant cultivars may also exhibit this purple colouring.

5. **Necrosis**

   Necrosis generally happens in later stages of a deficiency (probably following chlorosis) and causes the parts of the plant first affected by the deficiency to brown and die.

Since, nutrient deficiency symptoms are more or less categorised into 5 types, further evaluation of symptoms related to particular leaf patterns or locations on the plant are needed to diagnose nutrient specific deficiencies.

Stunted Growth

Chlorosis

Interveinal Chlorosis

Red Purplish Color

Necrosis

Fig. 1.2: *Some common deficiency symptoms*

Symptoms	Suspected Element									
	N	P	K	Mg	Fe	Cu	Zn	B	Mo	Mn
Yellowing of Younger Leaves					✓					✓
Yellowing of Middle Leaves									✓	
Yellowing of Older Leaves	✓		✓	✓			✓			
Yellowing Between Veins				✓						✓
Old Leaves Drop	✓									
Leaves Curl Over				✓						
Leaves Curl Under			✓			✓				
Younger Leaf Tips Burn								✓		
Older Leaf Tips Burn	✓						✓			
Young Leaves Wrinkle/Curl			✓				✓	✓		
Necrosis			✓	✓	✓		✓			✓
Leaf Growth Stunted	✓	✓								
Dark Green/Purple Leaf & Stems		✓								
Pale Green Leaf Color	✓							✓		
Molting							✓			
Spindly	✓									
Soft Stems	✓		✓							
Hard/Brittle Stems		✓	✓							
Growing Tips Die		✓						✓		
Stunted Root Growth		✓								
Wilting						✓				

Fig. 1.3: *Nutrient Deficiency Symptoms in Plants*
Courtesy:www.agrinfobank.com

Table 1.5 Nutrient Deficiency Symptoms in common Crop groups

CROP	NUTRIENT DEFICIENCY SYMPTOMS
NITROGEN	
Food Crops Wheat, maize, millets, pulses, cereals	Deficient plants are light green to yellow in color. Growth is stunted, Shoots are dwarfed Lower leaves are yellow which later dry. Tillering is restricted V- shaped portion of leaf margin remains green in maize
Cash Crops Sugarcane, Cotton, Tobacco	Plants have few lateral bunches, flowers and bolls in cotton Petioles turn yellow Leaves - particularly older leaves - become light green to yellow in color and plant growth is reduced. Leaf tips may become necrotic. Stalks are thin and stunted with reduced length between internodes. Reduced tillering and lower root mass.
Horticultural Crops Fruits and Vegetables	The usual symptoms are retarded growth and chlorosis Leaves are small in size having yellow green to light green appearance Old matured leaves are discolored Leaves may also become fragile and shed In few cases, twigs become slender and hard

PHOSPHORUS

Food Crops Wheat, maize, millets, pulses, cereals	Growth is retarded or stunted Foliage is dark green with purple tinge Leaves are small in size Stem develops a purple color

Cash Crops Sugarcane, Cotton, Tobacco	Immature growth
Foliage dark green
Leaves small
Color of the leaves in greenish blue or red - purple discoloration on tips and margins, narrow and somewhat reduce in length.
Reduction in length of sugarcane stalks,
Poor or no tillering.
Decreased shoot / root ratio with restricted root development.
Horticultural Crops Fruits and Vegetables	Growth is reduced
Older leaves are loose and develop into faded green to brown color which further turn necrotic
Younger leaves are stunted and dark green in color
Under surface of leaves develop reddish purple color

POTASH

Food Crops Wheat, maize, millets, pulses, cereals	Plants have stunted growth and leaf edge becomes mottled, dries up and is scorched
During the later stages, leaves are streaked with yellowish green color
Cash Crops Sugarcane, Cotton, Tobacco	Yellowish spots seen between veins
Leaf curls downwards
Yellowing and marginal drying of older leaves
Spots develop in between veins, margins and leaf tips
Development of slender stalks
Poor root growth with less member of root hairs.
Horticultural Crops Fruits and Vegetables	Growth retarded with shortened internodes
Leaves are crinkled and develop small necrotic dots
Twigs are stunted
Leaves and shoots shed extensively during blossom
Margins of older leaves develop yellowish brown and bronze color

MAGNESIUM

Food Crops
Wheat, maize, millets, pulses, cereals

Plants turn yellow and are dwarfed showing green patches
Main veins become pale green and turn deep yellow
Yellow streaks develop between parallel veins which finally dry up and die
Lowermost leaves show chlorosis at the tips and margins

Cash Crops
Sugarcane, Cotton, Tobacco

Chlorosis of leaves
Leaf margins and tips dry up
Lower leaves develop purple red color with green veins

Horticultural Crops
Fruits and Vegetables

Chlorosis appears between leaf veins of new leaves and then spreads to older leaves
Matured leaves show fawn colored patches. Affected leaves dry up.

ZINC

Food Crops
Wheat, maize, millets, pulses, cereals

Leaves become chlorotic and dry up
Light yellow streaks between veins followed by necrosis of older leaves

Cash Crops
Sugarcane, Cotton, Tobacco

Lower leaves are necrotic with brown grey spots
Lower leaves chlorotic at the tips and margins
Necrosis of the leaves followed by drying up of the veins
Shortened internodes

Horticultural Crops
Fruits and Vegetables

Leaves chlorotic, mottled, rosetted; sometimes with irregular spots
Plants short
Stem and petioles develop brown spots
Fruit quality and quantity affected
Developing fruits are small and malformed

IRON

Food Crops
Wheat, maize, millets, pulses, cereals

Plants have green leaves with yellow strips
The tops of plants and leaf tips die

Horticultural Crops
Fruits and Vegetables

Interveinal chlorosis
Fruit production is reduced and in cases of acute deficiency the trees die

COPPER

Food Crops
Wheat, maize, millets, pulses, cereals

Plant becomes pale green
Older leaves bend back sharply
Young leaves become light yellow in color and develop necrotic tips

Cash Crops
Sugarcane, Cotton, Tobacco

Brown spots throughout the leaf
During flowering, upper leaves wilt badly

Horticultural Crops
Fruits and Vegetables

Plants are stunted and become chlorotic
Leaves nay also become scorched and die back
Head formation is affected in cruciferous vegetables

MANGANESE

Food Crops
Wheat, maize, millets, pulses, cereals

Pale green leaves with faint chlorotic streaking or brown spots between veins
The growing point dies and tillering is absent in severely affected plants

Cash Crops
Sugarcane, Cotton, Tobacco

Stunted growth and entire plant becomes yellow
Whitish chlorotic streak develop on the leaves
Yellowing of leaf veins

Horticultural Crops Fruits and Vegetables	Interveinal chlorosis Flowering and fruit formation is interrupted

BORON

Food Crops Wheat, maize, millets, pulses, cereals	Interveinal chlorosis of old leaves Young leaves rolled Poor cob formation Abnormal tillering Ear formation checked
Cash Crops Sugarcane, Cotton, Tobacco	Growth is depressed or stunted Leaves chlorotic Necrotic spots on the stem and roots Flowering inhibited
Horticultural Crops Fruits and Vegetables	Leaves chlorotic/ discolored/ dwarfed or curled Dieback observed Stunted stems

MOLYBDENUM

Food Crops Wheat, maize, millets, pulses, cereals	Plants stunted with general yellowing Color of the foliage is dull pale green Flowering and grain formation are suppressed Ripening delayed
Cash Crops Sugarcane, Cotton, Tobacco	Crinkling and mottling of midstem leaves Leaves wither from the margins Stem height decreased Flowering delayed
Horticultural Crops Fruits and Vegetables	Leaf rough in texture, withered, curled and chlorotic Wilting of petioles Veins could turn purple Large interveinal yellow spots

NUTRIENT MOBILITY AND DEFICIENCY

The availability and mobility of essential nutrients within the soil and in the plant is variable and plays an important role in managing fertigation schedules.

Mobility in Soil:

The mobility of plant nutrients in the soil has a direct bearing on its uptake and its susceptibility to various processes such as fixation, leaching and run-off. For nutrients in soil to be available to plants, they must exist as ions - molecules with either a positive or negative charge. Ions are simply atoms or molecules with a charge, either positive or negative. Mobility is impacted by whether the charge of a nutrient is +(Cation) or -(Anion) and also the strength of the charge as well. Generally, positively charged ions (Cations) typical bind to soil while negatively charged ions (Anions) are repelled by soil particles and float freely. These Anions (-) which are repelled by the soil particles float freely in the water in soil. The anions want to disperse themselves to even concentrations so they move from higher concentrations to lower concentrations. Certain Anions do bind tightly to soil particles. The most infamous one being Phosphorus which notoriously locks up to soil particles, becoming unavailable to plants.

The composition of soil is diverse, and varies from soil to soil. While the strength of a soil's charge varies, soils generally maintain a negative charge with small pockets of positive charges intertwined. The strength of the soil charge is called the Cation Exchange Capacity which measures the number of cations that can be retained by soil particles. The higher the CEC, the more Cation nutrients that can be stored in the soil. So, higher CEC soils can become more nutrient rich.

Mobility in Plants

Once taken up by the roots, nutrients are transported to where they are needed, typically to growing points. Once the nutrient gets incorporated by the plant, some elements can be immobile while others can be remobilized.

Immobile elements essentially get locked in place and that is where they stay. Those that can be remobilized can leave their original location and move to areas of greater demand. Knowing which are mobile or immobile is helpful in diagnosing deficiency symptoms. Since immobile elements do not easily move within the plant, when deficiency symptoms occur they show up in new growth. When mobile elements become limiting, they can be scavenged from older growth and moved to where they are most needed, causing deficiency symptoms in older growth. Table 1.6 gives the nutrient mobility status in soils as well as in the plant.

Mobile nutrients are nutrients that are able to move out of older leaves to younger plant parts when supplies are inadequate. Mobile nutrients include N, P, K, Mg and molybdenum (Mo). Because these nutrients are mobile, visual deficiencies will first occur in the older or lower leaves and effects can be either localized or generalized.

Very mobile – Nitrogen, Phosphorus, Potassium, Magnesium (Deficiency symptoms appear first in older leaves and quickly spread throughout the plant)

Moderately mobile – Sulfur, Copper, Iron, Manganese, Molybdenum, Zinc (Deficiency symptoms first appear in new growth but do not readily translocate to old growth)

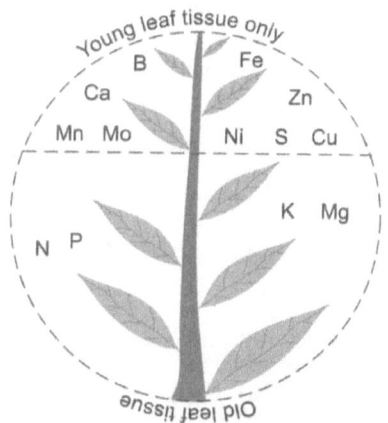

Fig. 1.4: Portion of the plant where nutrient deficiency symptoms are first observed. Courtesy: 4R Plant Nutrition Manual, IPNI 2012.

In contrast, immobile nutrients [B, calcium (Ca), Cu, Fe, Mn, S and Zn] cannot move from one plant part to another and deficiency symptoms will initially occur in the younger or upper leaves and be localized.

Immobile – Boron, Calcium (Calcium is very immobile)

Table 1.6 Nutrient mobility in Soil and Plants

Nutrient	Requirement	Non-ionic	Cation (+)	Anion (-)	Soil Mobility	Plant Mobility
Carbon	Basic	CO_2, H_2CO_3			N/A	
Hydrogen	Basic	H_2O	H^+	OH^-	N/A	
Oxygen	Basic	O_2			N/A	
Nitrogen	Primary Macronutrient		NH^{4+}	NO^{3-}	Mobile as $NO3^-$, immobile as $NH4^+$	Mobile
Phosphorus	Primary Macronutrient			HPO_4^{2-}, $H_2PO_4^-$	Immobile	Mobile
Potassium	Primary Macronutrient		K^+		Low Mobility	Mobile
Calcium	Secondary Macronutrient		Ca^{2+}		Low Mobility	Immobile
Magnesium	Secondary Macronutrient		Mg^{2+}		Immobile	Mobile
Sulfur	Secondary Macronutrient			SO^{4-}	Mobile	Mobile
Boron	Micronutrient	$H3BO3$		BO^{3-}	Mobile	Immobile
Copper	Micronutrient		Cu^{2+}		Immobile	Immobile
Iron	Micronutrient		Fe^{2+}, Fe^{3+}		Immobile	Immobile
Manganese	Micronutrient		Mn^{2+}		Mobile	Immobile
Zinc	Micronutrient		Zn^{2+}		Immobile	Immobile
Molybdenum	Micronutrient			MoO^{4-}	Low Mobility	Immobile
Chlorine	Micronutrient			Cl^-	Mobile	Mobile
Cobalt	Micronutrient		Co^{2+}		Low Mobility	Immobile
Nickel	Micronutrient		Ni^{2+}		Low Mobility	Mobile

1.3 DETECTING MICRONUTRIENT DEFICIENCIES: SOIL ANALYSIS AND TISSUE ANALYSIS

Plants need the 16 elements for proper growth and development. Nutrient deficiency occurs when the plant does not get a required nutrient in sufficient quantities. Without sufficient essential nutrients, plants will not grow well and show various symptoms to express the deficiency. A plant's sufficiency range is the range of nutrient amount necessary to meet the plant's nutritional needs and maximize growth (Fig. 1.6). Nutrient levels outside of a plant's sufficiency range cause overall crop growth and health to decline due to either a deficiency or toxicity.

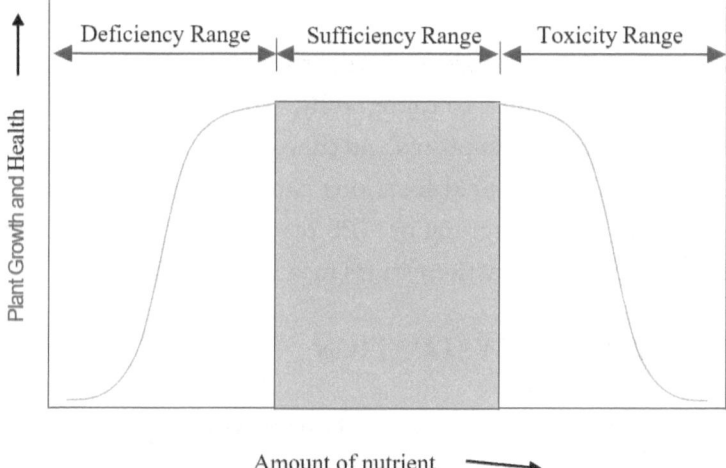

Fig. 1.5: *Relationship between plant growth and health and amount of nutrient available.*
(Brady and Weil, 1999)

Micronutrient deficiencies can be detected by visual symptoms on crops and by testing soils and plant tissues. Both soil testing and tissue analysis are quantitative estimations that compare soil or plant concentrations to the sufficiency range for a particular crop. Visual observation, on the other hand, is a qualitative assessment and is based on symptoms such as stunted growth or a yellowing of leaves occurring as a result of nutrient stressEnvironmental and physiological factors including, but not limited to, disease, drought, excess water, genetic abnormalities, herbicide and pesticide residues, insects, and soil compaction could give rise to certain visual symptoms that could be mistaken for nutrition deficiency.

Interpreting visual nutrient deficiency and toxicity symptoms in plants can be difficult and hence a confirmatory test either as a plant analysis or soil testing is needed. Plants may be nutrient deficient without showing visual clues. Visual observation is also limited by time. Between the time a plant is nutrient deficient (hidden hunger) and visual symptoms appear, crop health and productivity may be substantially reduced and corrective actions may or may not be effective. Therefore, regular soil or plant testing is recommended for the prevention and early diagnosis of nutrient stress.

If visual symptoms are observed, record which crop(s) are affected, their location with respect to topography, aspect, and soil conditions, a detailed description of symptoms and time of season that the symptoms first appeared. Affected field locations can be marked and monitored over time using either flagging or GPS readings. This information will be useful in preventing nutrient stress in subsequent years.

SOIL TESTING FOR EVALUATION

Soil testing and plant tissue testing are two tools for helping to understand fertility programs. Soil testing and plant analysis are complementary: however, they are extremely different.

Soil testing is the best tool for monitoring soil fertility levels and providing baseline information for cost-effective fertilization programs. Routine soil testing can identify nutrient deficiencies and inadequate soil pH conditions that may negatively affect forage production. Soil tests can also indicate nutrients that are present at adequate levels, providing the opportunity to eliminate unnecessary soil amendments. A major limitation associated with soil testing is that it typically accounts for the plant-available nutrient pool present in the surface (4 to 6 inches) soil layer. However, the subsoil can be an important source of water and nutrients, particularly in perennial crop systems. In addition, some nutrients are highly mobile in the soil and can easily leach into subsoil, resulting in nutrient accumulation at deeper soil depths

Getting a truly representative sample for each field should be the main focus in soil sampling. Commonly, samples are taken in random locations across the field, or from a smaller area of the field that is considered typical of the entire field (benchmark sampling). Geo-referencing the sampling locations using GPS is another tool that is currently used in soil testing today. All methods are effective if soil samples are taken from areas that best represent the field.

Unlike soil testing, plant tissue analysis can account for the plant-available nutrient pools present at multiple soil depths, including deeper horizons. Because of the extensive root system in some plants, plant analysis is a complement to the soil test to better assess the overall nutrient status of a perennial forage system, while revealing imbalances among nutrients that may affect crop production.

TISSUE TESTING FOR EVALUATION

Plant tissue tests are used for several reasons. A tissue test can be used for monitoring of high-value crop nutrition, such as the petiole analysis. Plant tissue analysis involves the determination of nutrient concentrations from a particular part or portion of a crop, at a specific time and/or stage of development. It can also be used for crop logging, which allows determination of critical levels, or for diagnosing crop fertility problems. Using tissue analysis to identify an imbalance or deficiency provides an opportunity to improve the fertility plan for the following year.

Using plant tissue analysis for this purpose works best when specific plant parts are sampled at appropriate growth stages. Most laboratories can analyze for various combinations of many plant macro- and micronutrients. Theoretically, a tissue test is used to establish the concentration of a nutrient in the plant tissue. That level is compared to a critical level required for plant growth. Recent efforts have shown that when plant tissue analysis was used in combination with soil testing, there was improved predictability of P and K availability to plants (Silveira *et.al.* 2011).

When used in conjunction with soil testing, tissue analysis will improve our diagnostic toolbox for developing nutrient management programs that predict when crops need additional nutrients, while avoiding unintended impacts of excess fertilization on the environment.

1.4 MICRONUTRIENT STATUS OF INDIAN SOILS

Micronutrients play an important role in Indian agriculture towards sustainable crop production. The importance of micronutrients need to be viewed in food systems context, as their inclusion in balanced fertilization schedule would optimize micronutrient supply and availability in the entire food consumption cycle (Takkar and Shukla, 2015). Proper crop nutrition management, macro as well as micro, is an extremely important precondition for healthy and vigorous crops.

An emerging concern, however, is widespread micronutrient deficiency in soils and crops. Most soils globally are witnessing multi-micronutrient deficiencies, dominated by zinc (Zn) and boron (B). This is adversely impacting yield and quality of crops. The scenario in India is no different and the Indian soils are primarily deficient in Zinc, Boron along with other micronutrients like Iron, Manganese, Copper etc. Indian soils are generally poor in fertility especially in micronutrients as these have consistently been mined away from their finite soil source due to continuous cultivation for a very long time without addition of micronutrient fertilizer resulting in emerging micronutrient deficiency. The major causes for micronutrient deficiencies are intensified agricultural practices, unbalanced fertilizer application including NPK, depletion of nutrients and no replenishment. In addition, green revolution led-increased demand of micronutrients by the high-yielding crop cultivars (especially rice and wheat) as well as adoption of intensive cropping practices, use of high-analysis fertilizers with low micronutrient content, decreased use of organic manures and crop residues, growing of crops in soils with low micronutrient reserves and other natural and anthropogenic factors adversely affecting phyto-availability of micronutrients and aggravated the situation (Takkar and Shukla, 2015).

The problem of micronutrients deficiency is widespread in horticulture crops as well. These crops suffer widely by zinc deficiency followed by Boron, Manganese, Copper, Iron (mostly induced) and Molybdenum deficiencies. Copper, Iron, and Magnesium are involved in various processes related to photosynthesis while Zinc, Copper, Iron, and Magnesium are associated with various enzyme systems. Boron is the only micronutrient not specifically associated with either photosynthesis or enzyme function, but it is associated with the carbohydrate chemistry and reproductive system of the plant. Thus, the importance of micronutrients in growth as well as physiological functions of horticultural crops is significant.

STATUS OF MICRONUTRIENTS IN INDIAN SOILS

The collection and analysis of more than 2.0 lakh soil samples from across 508 districts during 2011-2017 under the leadership of ICAR – Indian Institute of Soil Science, Bhopal, revealed that on an average 36.5% soils are deficient in Zn, 12.8 % in Fe, 7.1% in Mn, 4.2% in Cu and 23.2% in B.

Global positioning system (GPS) and geographical information system (GIS) based prepared district wise maps for various states of India help in understanding the level of micronutrient deficiency and toxicity and hence, formulating the remediation strategies for correcting the same.

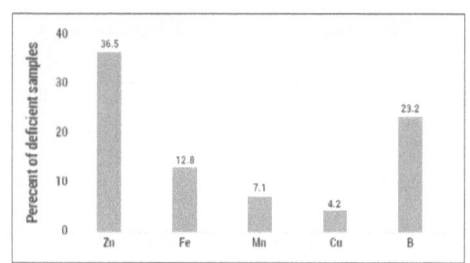

Fig. 1.6: Micronutrient deficiency in Indian soils (2017)
(Shukla, et al., 2019)

ZINC DEFICIENCY

The findings indicated that the most Zn deficient soils are the ones that are coarser in texture (sandy / loamy sand), high in pH (>8.5 or alkali/sodic soils) and or low in organic carbon (< 0.4%), or calcareous/high in $CaCO_3$ (> 0.5%) and intensively cultivated (Shukla *et.al.*, 2014). Hence among the Indian states, minimum Zinc deficiency of 9.6% was observed in Uttarakhand to as high as 75.3% in Rajasthan.

Low Zn deficiency has been recorded in soils having acidic pH as compared to soils having high pH; however, soils of northern India, except Rajasthan, also displayed medium deficiency range as compared to previous studies due to regular use of $ZnSO_4$ fertilizer (Sadana *et.al.*, 2010). On the other hand, increase in Zn deficiency in areas where low deficiency was recorded earlier has resulted from intensification of agricultural systems, faster depletion rate of available Zn pools (Shukla *et.al.*, 2016). Initially, the incidence of Zn deficiency was observed more in cereals, particularly rice and wheat belts of the country, but with passage of time, distribution of Zn deficiency has covered the whole country across the crops and cropping systems (Shukla and Tiwari, 2016).

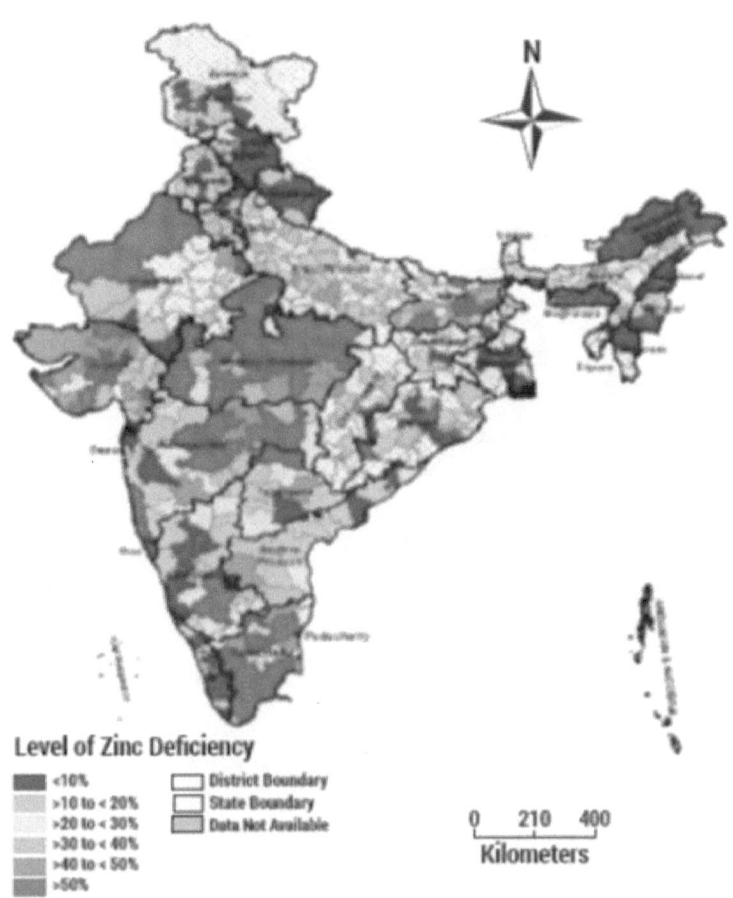

Fig. 1.7: *Zinc deficiency status in soils of India, 2011-17*

Iron Deficiency

Considering Fe deficiency, in India the problem of iron deficiency is mainly in calcareous and other alkaline soils having pH > 7.5. The availability of Fe gets reduced under drought or moisture stress condition due to conversion of Fe^{2+} iron to less available Fe^{3+} iron. The soils of north-eastern districts, Odisha and Kerala are reported to have Fe toxicity problem in rice paddies (Shukla, et.al. 2019).

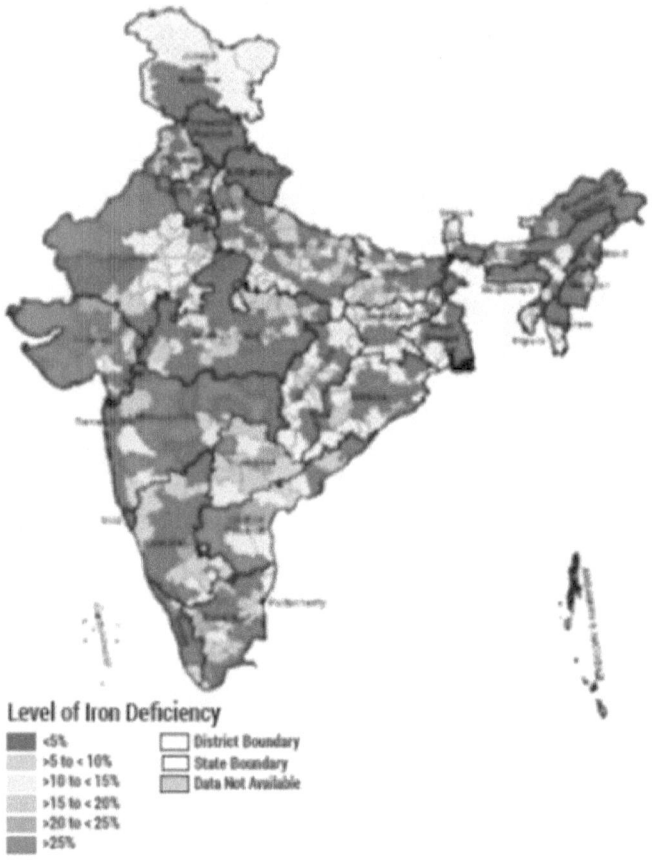

Fig. 1.8: *Iron deficiency status in soils of India, 2011-17*

Manganese Deficiency

Similar to Fe, Mn availability is also influenced by soil moisture, and affect the incidence and severity of Mn deficiency in crops grown with low moisture content. On the other hand, Mn is more mobile in imperfectly drained soils (water logged) and sometime exhibited Mn toxicity in rice grown in paddy fields under continuous submerged conditions.

Fig. 1.9: Manganese deficiency status in soils of India, 2011-17

Copper Deficiency

In India, Cu deficiency is not a such major concern showing deficiency in 4.2% soils only.

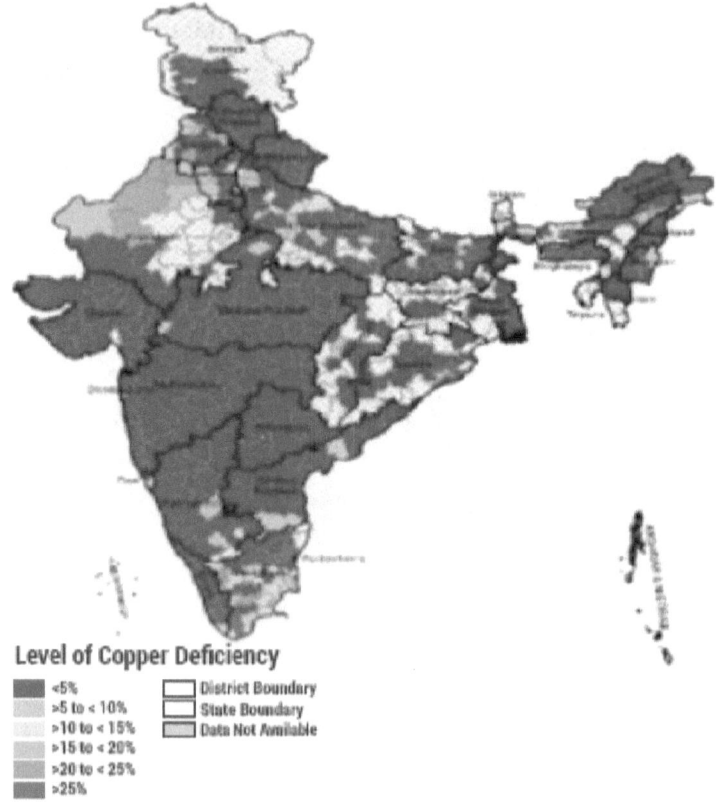

Fig. 1.10: Copper deficiency status in soils of India, 2011-17

Boron Deficiency

Boron deficiency is more common in highly calcareous soils of Bihar and Gujarat and acid soils of West Bengal, Odisha and Jharkhand.

In India, B deficiency has been recognised next to Zn. Availability of B to plants is governed by soil pH, $CaCO_3$ and organic matter contents, interactions of B with other nutrients, plant type and variety, and environmental factors. In general, B deficiency was higher in eastern region of the country and has resulted due to its excess leaching in sandy loam soils, alluvial and loess deposits.

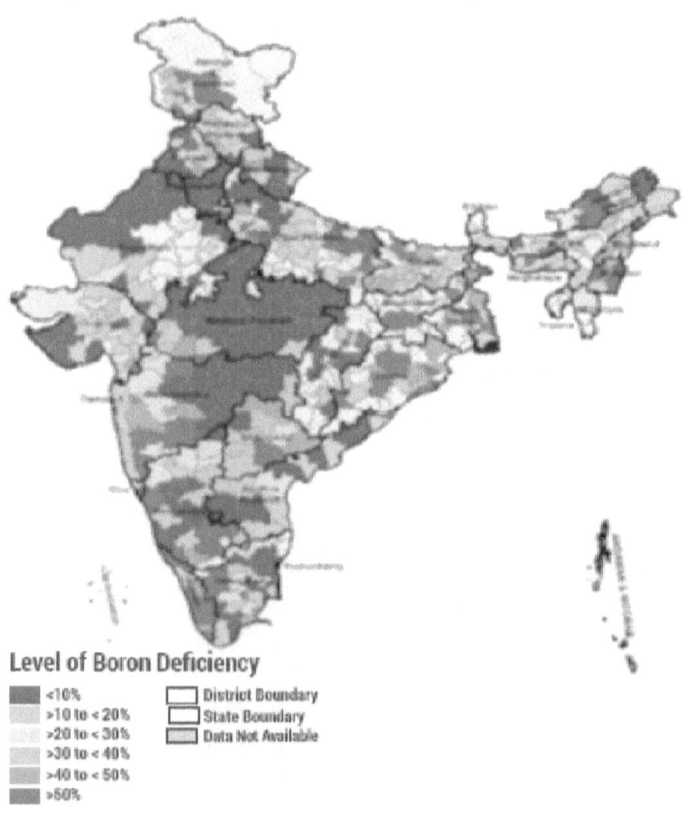

Fig. 1.11: Boron deficiency status in soils of India, 2011-17

Molybdenum Deficiency

Molybdenum is the least studied micronutrient in India. Molybdate anions (MoO_4^{2-}) are strongly adsorbed by soil minerals and colloids (at pH < 6.0) and sometimes also trapped due to formation of secondary minerals. Hydrous aluminium silicates may also fix Mo strongly. Soils formed from shale and granite parent materials had high Mo concentrations; whereas, those derived from sandstone, basalt and limestone had low Mo contents. Most of the soils are adequate in Mo but its deficiency is noticed in some acidic, sandy and leached soils. Molybdenum is most readily taken up by plants in soils with a pH above 7 and is relatively unavailable in acid soils. Thus, Mo deficiencies are most likely to occur on acid and severely leached soils and severely affecting mainly legumes, crucifer vegetables and oilseeds (Shukla, et.al., 2019).

Multinutrient Deficiency

With due course of time, multi-micro and secondary nutrients deficiencies have emerged in different areas of the country. Out of the 2 lakh samples analysed, about 9.9% of samples were deficient in sulphur and zinc, 8.3% samples in zinc and boron, 6.2 % in sulphur and boron, 5.8% in zinc and iron, 3.7% in sulphur and iron 3.3% in zinc and manganese, 2.8% in zinc and copper and 2.4% in iron and boron nutrient combinations. Three nutrient deficiencies like S+Zn+B, S+Zn+Fe and Zn+Fe+B were recorded in about 2.5, 1.9 and 1.1% soils, respectively. Four or more than four nutrient deficiencies were very less (less than 0.5%) in most of the states (Shukla, et.al, 2019).

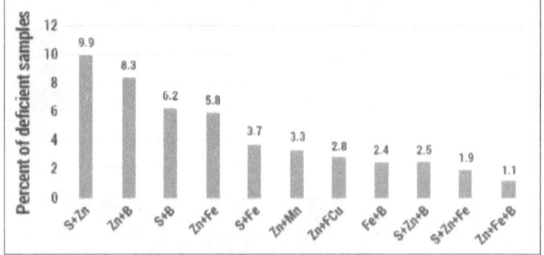

Fig. 1.12: Multimicronutrient and Secondary nutrient deficiency in soil (2017)
Shukla et.al., 2019

The results clearly indicate that single micronutrient deficiency is most predominant as compared to the combination of two or three or four or more than four elements simultaneously. Only the deficient micronutrient as revealed by the soil tests should be used to mitigate the deficiency and minimize environment pollution.

1.5 MICRONUTRIENT MANAGEMENT FOR HIGHER CROP PRODUCTION

Micronutrient management needs to be carried out based on the demand and supply of micronutrient in soil-plant system. It varies with crops, soil types, severity of deficiency, source, method, time, rates and frequency of application. While planning for replenishment of the micronutrients removed by the crop and/or depleted from soil through micronutrient management, important aspects like micronutrient requirements of the crops and cropping systems, ranges between their deficiencies and toxicities, low use efficiency of micronutrients, and residual availability etc. need to be considered (Shukla et.al., 2014). Multi-nutrient soil deficiency directly impacts crop productivity and indirectly contributes to malnutrition. The beneficial effects of balanced fertilization are better growth and productivity of crops which resulted in lower production costs, better profitability, and improved chances of producing a good yield under adverse climatic and soil conditions.

The optimum rates of Zn application varied with severity of Zn deficiency, soil types and nature of crops. Out of the several Zn sources evaluated for their efficacy under different soil-crop situations, $ZnSO_4 \cdot 7H_2O$ proved better or equal with other sources in correcting the Zn deficiency (Shukla et.al., 2009; Shukla and Behera, 2012). However, in some studies, chelated Zn proved more effective than $ZnSO_4 \cdot 7H_2O$ for maize and rice. As the efficiency of soil applied Zn is very low (2-5%), efforts have been made to develop efficient and inexpensive methods of Zn application. Application of Zn to soil through broadcast and mixed or its band placement below the seed proved superior to top dressing. Other Zn application methods are side-dressing or side-banding, foliar application of 0.5 to 2.0% $ZnSO_4·7H_2O$ solution and soaking or coating of seeds in Zn solution (Shukla, *et.al.*, 2019).

Iron chlorosis is generally observed in upland crops especially rice, sorghum, groundnut, sugarcane, chickpea grown in highly calcareous soils, compact soil with restricted aeration, soils with low in Fe and high in P and bicarbonate. The rates of soil application of Fe were very high, because of rapid rate of oxidation of Fe^{2+} to Fe^{3+} and as such were uneconomical. Ferrous sulphate, (19-20.5% Fe) is the major source used for managing Fe deficiency in the country. However, Fe-EDTA (9-12% Fe), Fe-EDDHA (6.0% Fe), pyrite, biotite, and organic manures (FYM 0.15% Fe), poultry and pig manure (0.16% Fe), sewage sludge have also been used as sources of Fe to correct its deficiency in crops. For horticultural crops foliar spray of Fe is recommended, which have been more effective and efficient than soil application in correcting Fe-chlorosis in tomato, chilli, groundnut and sugarcane.

Severe Mn deficiency is difficult to manage with soil application due to oxidation of soil-applied Mn, especially in high pH soils. Foliar application of Mn is an immediate effective measure to combat Mn deficiency in wheat though it has to be applied every year.

Copper deficiency in Indian soils is very less. Either soil or foliar application of Cu to soybean-wheat cropping system proved equally effective in correcting its deficiency and gave significant response of 0.2 t/ha with soil application of 5.0 kg Cu/ha to the first crop. Foliar spray of 0.2% $CuSO_4$ solution increased soybean grain yield from 2.18 to 2.35 t/ha. Residual effect of soil applied Cu on the following wheat was non-significant (Shukla *et. al.*, 2019).

Boron deficiency is one of the serious nutritional problems limiting crop production in acid and calcareous soils. Soil application of borax or sodium tetra-borate decahydrate ($Na_2B_4O_7·10H_2O$, fertilizer grade, 10.5% B) is commonly used to correct its deficiency. Boric acid (H_3BO_3,

17% B), solubor ($Na_2B_4O_7 \cdot 5H_2O$ + $Na_2B_4O_7 \cdot 10H_2O$ fertilizer grade, 19% B) are mostly used as foliar spray. However, rates of B application for achieving sustainable optimum productivity varies with crop, season and type of soil. By and large, soil application of B is a better method of its management than the foliar and seed soaking.

Micronutrient management depending upon crops, soil types, severity of deficiency, source, method, time, rates and frequency of application needs to be undertaken for sustainable agricultural production and maintenance of human health.

Micronutrient Fertilisers

2.1 FERTILISER CONTROL ORDER

Fertilisers, being the most essential input needed for quality growth of crops, it is necessary that the farmers get them in the right form without any loss in quality. To ensure adequate availability of right quality of fertilisers at right time and at right price to farmers, the Fertiliser was declared as an Essential Commodity and Fertiliser Control Order (FCO) was promulgated under Section 3 of Essential Commodities Act,1955 to regulate, trade, price, quality and distribution of fertilisers in the country.

The FCO provides for compulsory registration of fertiliser manufacturers, importers and dealers, specification of all fertilisers manufactured/ imported and sold in the country, regulation on manufacture of fertiliser mixtures, packing and marking on the fertiliser bags, appointment of enforcement agencies, setting up of quality control laboratories and prohibition on manufacture/import and sale of non-standard/spurious/ adulterated fertilisers.

The Joint Secretary (INM) in the Ministry of Agriculture and Farmers Welfare, is the Controller of Fertilisers and the overall authority in the matter at the national level. The analytical methodology and its evaluation, monitoring and technical training, etc. is the responsibility of the Ministry's Central Fertiliser Quality Control and Training institute (CFQC&TI). Accordingly, there are 74 notified Fertiliser Quality Control Laboratories in the Country which includes 4 set up by

Central Government as CFQC&TI, Faridabad and its three Regional Laboratories.

The order also provides for cancellation of authorization letter/registration certificates of dealers and mixture manufacturers and also imprisonment from 3 months to 7 years with fine to offenders under ECA. the FCO offence has also been declared as cognizable.

The FCO today includes all the common fertilisers containing N, P, K, S, Ca, Mg, single micronutrient products, salts as well as chelates, a number of specialty products such as 100% water soluble fertilisers (primarily meant for fertigation), fortified fertilisers, liquid fertilisers and also neem coated urea. Specifications of fertiliser mixtures including multi micronutrient products have been left to individual state departments of agriculture.

The enforcement of this Order has primarily been entrusted to State Governments. The Central Government provides training facilities and technical guidance to States and supplements their efforts through random inspection of manufacturing units and their distribution network through the Inspectors

MICRONUTRIENT FERTILISERS IN THE FCO

Micronutrient fertilisers include inorganic salts, chelated fertilisers and multi micronutrient formulations. In addition, several straight and complex fertilisers are also being fortified with micronutrients, particularly B and Zn.

Zinc Fertilisers:

Zinc fertilisers, both in the sulphate as well as the chelate are recognised by the FCO. Zinc Sulphate, heptahydrate (21% Zn and 10% S) and monohydrate (33% Zn and 15%S). Both are agronomically equally effective on per kg Zn basis.

Zinc- EDTA (12% Zn) / Zinc-HEDP (17% Zn) are also registered in the FCO and can be applied either through soil or as a foliar spray.

Boron Fertilisers:

The two common boron containing fertilisers are Borax (sodium borate containing 10.5% B) and Boric acid (17% B). Di sodium octaborate tetrahydrate containing 20% B is among the more concentrated sources of B.

Copper Fertilisers:

Copper sulphate pentahydrate (Cu 24%and S 12%)and Copper sulphate monohydrate (Cu 35%) are the two sources of water soluble and plant available copper mentioned in the FCO.

Iron Fertilisers:

Ferrous sulphate (Fe 19% and S 15%) and Fe-EDTA (Fe 12%) are the iron fertilisers mentioned in the FCO

Manganese Fertilisers:

Manganese Sulphate (Mn 30.5% and S17%) is mentioned in the FCO as Mn containing fertiliser.

Molybdenum Fertiliser:

Ammonium molybdate contains 52%Mo and some Nitrogen. Mo is in water soluble and plant available form.

2.2 STATE SPECIFIC GRADES

Table 2.1 Recommended grade as per Government of Andhra Pradesh

Formulation No. (1)	Crop to Which Grades Recommended (2)	Contents in Standards % Elements					
		Fe (3)	Mn (4)	Zn (5)	Cu (6)	Mo (7)	B (8)
1	Upland Paddy, Groundnut & Sugarcane	6	1.5	5	0	0	0
2	Oilseeds & Pulses	2	2	6	0	0	0
3	Low Land Paddy	Only Soil Application of Zinc Sulphate					
4	Citrus	4	3	6	1	0.05	2
5	Grapes	1	1	5	0	0	0.5
6	Vegetables & Cotton	2	2	5	0	0	0.5
7	Soil Application	0.5	0.5	6	0	0	0

Table 2.2 Recommended grade as per Government of Assam

Sr. No	Name of the Mixture	Specifications Nutrient content in % wt.by wt.					Remarks
		Zn	B	Mg	Cu	Mo	
Grade-1	Foliar Powder or Liquid	5.0	2.5	2.0	0.5	0.02	1. PH not less than 3.5 2. Solubility for powder 94% and Liquid 99%
Grade-2	Solid Mixture for Soil Application	7.0	5.0	0.0	0.5	0.03	Minimum Solubility 90%

Table 2.3 Recommended grade as per Government of Bihar

Sr.No	Name of the Micronutrient Mixture	Type of Application	Fe (Chelated)	Mn	Zn (Inorganic)	Zn (Chelated)	Cu	Mo	B	
1	Calcareous soil South Bihar Plain	Foliar	3.00	0.50	8.00	5.00	0.20	0.01	1.50	Solubility of foliar 90%. PH not
		Soil	4.00	0.50	10.00	5.00	0.20	0.02	2.00	less than 3.5
2	North Bihar Non recent Alluvian Tarrai Sub Himalayan	Foliar	0.50	0.20	5.00	3.00	0.50	0.02	0.50	
		Soil	1.00	0.20	8.00	4.00	1.00	0.03	1.00	
3	Tal land and Acid soil 90% Calair pH not less than 5.5	Foliar	-	-	2.00	1.00	-	0.05	1.50	Foliar spray 0.50 to 14, 2-3
		Soil	-	-	4.00	2.00	-	0.05	2.00	times soil 20-30 kg/hect

Table 2.4 Recommended grade as per Government of Chhattisgarh Notification No 4919

Notified Grade No.	Name of the Micronutrient Mixture	Nutrient Content % W/W					
		Zn	B	Mo	Mn	Fe	Cu
*1	Soil Application	6.00	1.20	-	4.00	3.00	0.80
*2	Soil Application	5.00	1.60	-	3.00	5.00	1.50
*3	Foliar Application	5.00	2.00	0.30	3.00	4.00	1.00

Table 2.5 Recommended grade as per Government of Chhattisgarh Notification No 4022

Notified Grade No.	Name of the Mixture	Nutrient content % W/W				
		Zn Zinc	Mn Manganese	Bo Boron	Mo Molybdenum	Fe Iron
1	FOR FOLIAR SPRAY Inorganic Foliar Spray in powder/liquid form	3.00	-	0.10	0.02	-
2	Inorganic Foliar Spray in powder/liquid form	3.00	0.50	0.10	0.02	-
3	Inorganic Foliar Spray in powder/liquid form	3.00	0.50	0.10	-	3

Table 2.6 Recommended grade as per Government of Gujarat

Sr.No	Grade	Specifications Nutrient content in % wt. by wt.				
		Fe	Mn	Zn	Cu	B
1.	Foliar Spray 1.Normal	2.0	0.5	4.0	0.3	0.5
	2.Fe Deficient	6.0	1.0	4.0	0.3	0.5
	3. Zn Deficient	2.0	0.5	8.0	0.5	0.5
	4. Fe & Zn Deficient	4.0	1.0	6.0	0.5	0.5
2.	Soil Application 1.Normal grade	2.0	0.5	5.0	0.2	0.5

Table 2.7 Recommended grade as per Government of Himachal Pradesh

Sr.No.	Crops, Mode of Application & Grades	Micronutrient mixture grades for Himachal Pradesh Contents Percentage							
		Zn	Cu	Mn	Fe	B	Mo	Mg	Ca
1	Temperate Fruit Crops Foliar Spray (Liquid/Powder)	*2.0	*0.5	*0.5	-	1	0.2	-	1
2	Sub-tropical fruits & vegetables/foliar spray (Liquid/Powder)	5.0	0.5	2.0	2.0	1.5	0.1	1.0	-
3	Vegetables & all other crops (inorganic mix. Soil application)	5.0	0.5	2.0	2.0	1.5	0.1	1.0	-

Table 2.8 Recommended grade as per Government of Jharkhand

Sr. No.	Type of Land	Type of Application	Micronutrients(%)				Remarks
			B	Zn	Mo	Cu	
1.	Upper/Medium Land (Vegetables, Oil Seeds, Pulses, Maize & Other Crop of Upper Land)	Soil Foliar	2.00 1.00	2.00 1.00	0.05 0.02	0.5 0.2	All grades will contain filler A) Calcium & Magnesium to be used as filler B) For Soil supplication Dolomite will be used as filler C) It is advisable to use Zinc & Copper in form of Sulphate for Micronutrient Mixture Fertiliser D) Solubility for Foliar spray should be 90% E) The packing size of Micronutrient Mixture of Fertiliser will be as under 1. For Soil application from minimum 250gms to 5kg 2. For Foliar spray from minimum 50gm or 50ml to 1kg/1 ltr
2.	Lower Land (Paddy & Wheat)	Soil Foliar	1.00 0.5	4.00 2.00	0.00 0.00	0.5 0.2	

Table 2.9 Recommended grade as per Government of Karnataka

Sr. No.	Soil Type	Specifications Nutrient content in % wt. by wt.			
		Zn	Fe	Mn	B
1	Foliar Spray	3.0	2.0	1.0	0.5
	1. Black Soils with alkaline reaction				
	2. Red & Laterite Soils with acidic reaction	3.0	0.5	0.2	0.5
	3. Hilly & Coastal region with acidic ph	3.0	-	-	0.5
2	Soil Application	10.0	5.0	2.0	0.3
	1. Black Soils with alkaline reaction				
	2. Red & Laterite Soils, hilly & coastal region with acidic reaction	10.0	-	-	0.5

Table 2.10 Recommended grade as per Government of Madhya Pradesh

Sr.No	Name of the Mixture	Specifications nutrient content in % wt. by wt.			
		Zn	B	Mn	Cu
1.	Inorganic Foliar Spray	FOLIAR SPRAY			
		3.0	0.1	0.5	0.0
2.	Inorganic Percentage	FOR SOIL APPLICATION			
		5.0	0.5	1.0	0.5

Table 2.11 Recommended grade as per Government of Maharashtra

Sr. No	Type of Fertiliser	Grade of Fertiliser	Status
1.	Micronutrient Mixture Fertiliser	Grade-1. INORGANIC CHELATED SOIL APPLICATION Zn-5%, Cu-0.5%, Fe- 2%, Mn- 1%, B-1%	Active
2.	Micronutrient Mixture Fertiliser	Grade-2. INORGANIC / CHELATED FOILAR APPLICATION Zn-3%, Cu-1%, Fe-2.5%, Mn-1%, B-0.5%, Mo- 0.10%	Active
3.	Micronutrient Mixture Fertiliser	Grade-3. SOIL APPLICATION (FOR ACID SOIL) Zn-5%, Cu-0.5%, Fe-3%, Mn-2%, B-0.5%, Mo-0%	Active
4.	Micronutrient Mixture Fertiliser	Grade-4. SOIL APPLICATION FOR ALKALINE SOIL Zn-6%, Cu-0.8%, Fe-4%, Mn-3%, B-0.8%, Mo-0%	Active
5.	Micronutrient Mixture Fertiliser	Grade-5. FOLIAR APPLICATION FOR ACID SOIL Zn-3%, Cu-0.5%, Fe-2%, Mn-1%, B-0.8%, Mo-0.10%	Active

6.	Micronutrient Mixture Fertiliser	Grade-6. FOLIAR APPLICATION FOR ALKALINE SOIL Zn-6%, Cu-0.8%, Fe-4%, Mn-3%, B-1.2%, Mo-0.10%	Active
7.	Micronutrient Mixture Fertiliser	Grade-7. SOIL APPLICATION FOR ACID SOIL (For Zinc Deficiency) Zn-10.0%, Cu-0%, Fe-0%, Mn-0%, B-0.5%, Mo-0%	Active
8.	Micronutrient Mixture Fertiliser	Grade-8. SOIL APPLICATION FOR ALKALINE SOIL (For Ferrous Zinc Deficiency) Zn-10.0%, Cu-0%, Fe-5.0%, Mn-0%, B-0.0%, Mo-0%	Active
9.	Micronutrient Mixture Fertiliser	Grade-9. FOLIAR APPLICATION FOR ACID SOIL (For Zinc Deficiency) Zn-3%, Cu-0%, Fe-0%, Mn-0%, B-0.5%, Mo-0%	Active
10.	Micronutrient Mixture Fertiliser	Grade-10. FOLIAR APPLICATION FOR ALKALINE SOIL (For Pulse Crop) Zn-5.0%, Cu-0%, Fe-2.5%, Mn-0%, B-0.50%, Mo-0.10%	Active
11.	Micronutrient Mixture Fertiliser	Grade-11. FOLIAR APPLICATION FOR ALKALINE SOIL (For Cotton Crop) Zn-5.0%, Cu-0%, Fe-2.5%,. Mn-0%, B-0.50%, Mo-0%	Active

Table 2.12 Recommended grade as per Government of Orissa

Grade No	Soil/Foilar	Crops/Soils for which recommended	Micronutrients (%)							Remarks
			Zn	Fe	Mn	Cu	B	Mo	Others	
1	Foliar	All	3	0	0.2	0.1	0.2	0.005	0.5mg	Preference Cotton
2	Foliar	All	6	0	0	0	0.4	0.005		Preference Maize
3	Foliar	Fruit Crops	7	0.5	2	1	0.6	0.005		
4	Foliar		3	0.5	1	0.05	0.5	0.005		
5	Soil Application	Acid Soils	5				0.5		1.0mg	
6	Soil Application	Alkaline Soils	6	0.5	0.5	0.5	0.5			
7	Soil Application	Maize	10				0.5		1.0mg	

Fe percent should not exceed by 10% of the prescribed value, higher concentration of nutrients (excepting Fe) and addition of other nutrients N,P,K,Ca,Mg & S are allowed. In case of B & Mg. The prescribed value should not exceed 10%

Table 2.13 Recommended grade as per Government of Tamil Nadu

Sr.No.	Crop	M.N.Mix.No.	Specifications Nutrient content in % wt. by wt.						
			Fe	Mn	Zn	B	Cu	Mo	Mg
1	Groundnut	I	3.80	1.46	4.20	1.57	0.00	0.07	0.0
2	Millet	II	5.70	9.15	2.31	0.52	1.00	0.00	0.0
3	Cotton	III	3.80	2.99	3.15	3.15	1.25	0.07	0.0
4	Coconut	IV	3.80	4.80	5.00	1.60	0.50	0.00	0.0
5	Citrus	V	2.60	4.20	7.06	0.60	2.00	0.05	0.0
6	Vegetables	VI	7.60	1.22	1.68	2.48	1.0	0.14	0.00
7	Pulses	VII	3.80	6.10	4.00	2.10	0.0	0.35	0.00
8	Sugarcane (Foliar)	VIII	3.00	1.50	5.50	0.80	0.10	0.10	1.30
9	Cotton (Foliar)	IX	2.00	1.00	2.50	0.10	0.10	0.01	4.00
10	Paddy (Foliar)	X	1.00	0.50	5.00	0.05	0.35	0.00	6.00
11	Paddy(Basal)	XI	1.60	0.30	3.00	0.20	0.40	0.00	4.00
12	Sugarcane (Basal)	XII	4.75	0.35	6.00	0.20	0.20	0.00	1.25
13	Basal Application	XIII	3.04	3.66	4.20	2.10	1.00	0.00	0.00
14		XIV	3.81	4.58	7.35	2.10	2.50	0.00	0.00

Table 2.14 Recommended grade as per Government of Rajasthan

Notified Grade No.	Grade	Contents in % by weight minimum					
		Fe	Mn	Zn	Cu	Mo	B
1	Inorganic foliar Spray for all Crops & Soils	2.0	2.0	5.0	0.5	0.05	0.5
2	Inorganic Soil Application	2.0	2.0	5.0	0.5	0.05	0.5
3	Partly Chelated, Foliar Spray	2.0	2.0	5.0	0.5	0.05	0.5
4	Partly Chelated, Soil Application	2.0	2.0	5.0	0.5	0.05	0.5

Table 2.15 Recommended grade as per Government of Telangana

Sr.No.	Crops to which grades recommended	Content in Standards % elements					
		Fe	Mn	Zn	Cu	Mo	B
1.	Upland paddy, Groundnut & Sugarcane	6	1.5	5	0	0	0
2	Oil seeds & Pulses	2	2	6	0	0	0
3	Low Land Paddy	Only Soil application of Zinc Sulphate					
4	Citrus	4	3	6	1	0.05	2
5	Grapes	1	1	5	0	0	0.5
6	Vegetables & Cotton	2	2	5	0	0	0.5
7	Soil Application	0.5	0.5	6	0	0	0

Table 2.16 Recommended grade as per Government of Uttaranchal

Notified Grade No.	Grade	Contents in % by weight minimum				
		Zn	Fe	Mn	Cu	B
1	Inorganic Foliar Spray for Wheat Paddy & other Cereals	4.0	2.0	0.5	0.5	-
2	Inorganic Foliar Spray for vegetables & crops	3.0	1.5	-	0.5	0.5
3	Inorganic for all cereals	6.0	3.0	1.5	0.5	-

Table 2.17 Recommended grade as per Government of Uttar Pradesh

Notified Grades No.	Description of grade	Content in % by Wt. Minimum					
		Zn	Fe	Mn	Cu	B	PB
1	Inorganic foliar spray (For Wheat, Paddy & other Cereals)	4	2	0.5	0.5	-	0.003
2	Inorganic foliar spray (For vegetable crops)	3	1.5	-	0.5	0.5	0.003
3	Inorganic Soil Application (For all cereal crops)	6	3	1.5	0.5	-	0.003
4	Inorganic Foliar spray (For all crops)	6	3	2	1	1	0.003
5	Inorganic Soil application (For all crops)	10	5	2	1	1	0.003

Table 2.18 Recommended grade as per Government of West Bengal

Grade No.	Name of the mixture	Specifications Nutrient content in % wt. by wt.						Remarks
		B	Zn	Mo	Cu	Mn	Fe	
1	FOR FOLIAR SPRAY Inorganic Foliar spray in powder form	1.0	7.5	0.5	0.0	0.0	0.0	Vegetables, Oilseeds, Pulses, Wheat, Jute & Flowers
2	Inorganic foliar spray in Liquid form	0.5	5.0	0.25	0.0	0.0	0.0	Vegetables, Oilseeds, Pulses, Wheat, Jute & Flowers
3	Inorganic Foliar spray in powder form	0.5	8.0	0.0	0.0	0.0	0.0	Vegetables, Oilseeds, Pulses, Wheat, and Potato
4	Inorganic foliar spray in Liquid form	0.5	5.0	0.0	0.0	0.0	0.0	Vegetables, Oilseeds, Pulses, Wheat, and Potato
5	Inorganic foliar spray in Powder form	1.0	5.3	0.1	2.4	5.0	0.0	Fruit crops including Citrus & pineapple

1	FOR SOIL APPLICATION Inorganic soil application in Powder form	0.7	3.6	0.0	0.8	4.3	6.6	Vegetables, Wheat, Oilseeds, Pulses & Paddy under upland situation & Citrus
2	Inorganic soil application in powder form	0.7	3.6	0.0	0.8	0.0	0.0	Kharif Paddy under waterlogged situation, Jute, Vegetables, Wheat, Pulses, Oilseeds & Boro Paddy

2.3 ECONOMICS OF FERTILISER USE

Agronomic, economic, and managerial factors influence the use of fertiliser and must be considered together in relation to the crop to be fertilized and in relation to succeeding crops and the farm as a whole. Other factors also affect the response to fertilization. Different combinations of things such as density of soil, water supply, variety of crop, or degree of control of insects and diseases affect the results a farmer hopes to get from fertilisers.

Fertiliser application is worthwhile only when the increase in crop yield it produces also results in higher income for the farmer. Since the farmer wants to know how much profit to expect if he buys extra fertiliser, the tests are interpreted as an estimation of increased crop production that will result from nutrient additions. Cost of nutrients must be balanced against value of crop or even against alternative procedures, such as investing the money in something else with a greater potential return. Economics of fertiliser use tells us several things such as

- Whether or not the fertiliser use is profitable
- The amount of profit or loss
- The optimum or profit maximising dose to be applied
- The rate of return (value of crop per rupee spent on fertiliser) and
- The effect of different management practices and other factors or decisions, post-harvest losses etc. on the profitability of using fertiliser.

Crop yield increases with increasing doses of fertilisers until it reaches an optimum level. The extra yield obtained by the last fertiliser increment, termed as marginal response, tells us about the effectiveness of the final increment and help to arrive at the most profitable rate of fertiliser use or the cut-off point beyond which it is not worth adding more.

DETAILED ECONOMIC ANALYSIS

Detailed economic analysis for fertiliser use involves a farmer capturing information on several items on the expenditure and income side. Ideally this data should be quantified on a per acre/hectare basis reviewed every season.

On the Expenditure Side the farmer has the following expenses:

- Net Cost of fertiliser, pesticides and other agri inputs
- Transport
- Application / Labour Costs
- Direct / Indirect marketing costs

Similarly, on the Income Side the following are the revenue streams

- Sale proceeds from main crop
- Sale proceeds from by products, if any

The farmers also have to be provided with the following information:

1. **Cost Optimization**:

   Every farm process needs to be analysed with regard to labour requirement, mechanisation and substitutes of available agri inputs. Cost optimization at the farm level cannot disregard agronomic requirements of the crops. Looking at only commercial aspects/ cost revenue may be detrimental to farm productivity and hence balancing cost benefit is crucial.

2. **Subsidies**:

   Farmers have to be aware of Government subsidies and schemes which are dynamic and traditional change based on Central and State Government rules. The subsidies would reduce capital expenditure cost while optimising farm output.

3. **Substituting traditional fertilisers with Innovative new options**:

   It is essential to educate farmers on new fertiliser options and their consequent agri economic benefits. Shifting from traditional straight fertilisers to customized new age formulations involves demand creation using soil testing, demos, sampling and abundant capital. Training farmers using cost benefit ratios will lead to medium to long term results.

4. **Calculating Net Returns**:

   The gross returns from farm operations includes the total sale proceeds from the crop and the by-products, if any. From this Gross Returns all expenditure (as listed above) need to be deducted and the Net Returns percentage may be calculated. The expenditures should not exclude any inputs and incidental costs since the total input and output ratio for the farm operation needs to be optimized.

Based on the above calculation Net Return per acre/hectare may be calculated and reviewed every season. This crucial data combined and when correlated with the crop/soil conditions and input usage will lead to significant benchmark and enable farmers to make informed choices.

CALCULATION OF COST BENEFIT RATIOS

Following a series of trials conducted at the NHRDF, the table below provides a template to calculate cost benefit ratios on a protocol based research institution trials conducted on Tomato crop.

Using data extracted from the Trial Reports (attached and annexed as Trial Reports No19 and 20). The economics of fertiliser usage may be evaluated as under:

Table 2.19 Cost- Benefit analysis of chelated micronutrients

		Amino acid chelated multi micronutrients (1)	Amino acid chelated zinc through foliar spray (2)	Amino acid chelated zinc through drip (3)	EDTA chelated Zinc basal application (4)
Average market rate of tomato (Rs/qtl) Month: October Year: 2012		1030	1030	1030	1030
Yield difference (qtl/ha)		83.1	16.96	65.45	36.37
Return from extra yield (Rs)		85593	17468.8	67413	37461
Extra cost incurred on product (Rs)	MRP	2157	2456	275	22.5
	20%+dealer rate	1403	1445	412	357.5
Net return (Rs)	MRP	83436	15013	67688	37483.5
Estimated Farmer price (Rs)	20%+dealer rate	84190	16024	67825	37818.5
Cost Benefit Ratio	MRP	01:39	01:06	01:50	01:24
	20%+dealer rate		01:11	0.1	01:45

For e.g. In Trial '1' above, the yield increase reported from the usage of Amino acid based chelated multi micronutrients worked out to 83.10 per quintal per hectare, considering the average price of Tomato (as of month& year Oct-2012) of Rs.1030 per quintal. The return from this extra yield was Rs. 85,593. In order to secure this extra yield, the farmer incurred an additional cost of Rs.1403 (Farm Price estimated as

Dealer Rate + 20%) the Net Return therefore was Rs. 85,593 – 1403 = 84,190. The benefit cost ratio was Rs. 84190:1403 = 60:1. This assumes all other inputs on control and experimental plot is considered and the only difference being the Rs.1403 spent on Amino Acid Chelated Multi Micronutrients.

Similar cost benefit ratios may be calculated for the remaining trials in the above table with benefit cost ratio ranging from 6:1 to 60:1 (see table above).

The above calculation indicated the tremendous potential to increase farm income significantly by adopting integrated nutrient management practices and following innovative new age fertiliser schedules.

Working out the economics of fertiliser use is not a onetime exercise, but is a continuous drill and has to be repeated whenever there is a change in the price of the fertiliser, the produce or the response ratios. With changes in technology, release of new products, improved crop varieties and adoption of better application techniques response ratios can also change. Profitability of fertiliser use is directly and positively impacted by Fertiliser use efficiency. Any factor, decision or event which improves the yield increase per kg nutrient applied will improve the profitability of fertiliser use.

Agronomic and economic principles operate together in establishing the rate needed to get a specified return per rupee spent for fertiliser. Yield response that shows the rate of decreasing increments in the yield is the foundation for economic interpretation. The principle that marginal return equals marginal cost at the point of highest profit is general. But, as we have seen, many natural factors and many management practices influence yield response and modify the operation of the principle.

2.4 IMPACT OF FERTILISER APPLICATION ON CROP YIELDS

Close to 55% increase in agricultural production in recent years has been credited to fertiliser alone (FAO report on Agriculture towards 2000). A 50% contribution of fertilisers to foodgrain production means that the food for one out of every two persons added to the population comes from fertiliser. Thousands of trials have been conducted by scientists on the farmers field to estimate the yield increase brought about by fertilisers.

Often a soil needs more than one nutrient, balanced nutrient application produces a bigger yield as compared to the use of a single nutrient. Various stake holders, right from policy makers to farmers consider that balanced fertiliser use only refers to the application of NPK that too, in the classical ratio of 4:2:1. However, current trends do not always favour this. Large scale deficiencies of other plant nutrients have emerged in rev cent years which have made it necessary to include all 16 essential elements as part of the balanced fertiliser schedule and use. Currently, soil deficiencies are not limited to only a single nutrient and these must be corrected if good yields are to be achieved.

Effect of multi-micronutrient application on crop yields:

EFFECT OF ZINC ON CROP YIELD

Zinc is an essential micronutrient that has significant role in basic plant metabolic processes and enhances the growth, yield and quality by stimulating chlorophyll production, photosynthetic activity, nutrient uptake and protein biosynthesis.

More information is available on Zinc as compared to any other nutrient. Zinc is in fact considered as a micronutrient of macro importance. Since Zn deficiencies are widespread, impact of Zn application on increasing crop yields has been recorded on most crops.

Rice and wheat accounted for 89% of the total trials conducted. Zinc application increased paddy yield by 200-500kg/ha in 39% of the trials and by 500-1000 kg/ha in 23% trials in rice and 22% trials with wheat. Table 2.2 gives the average crop responses to Zinc in field experiments (HLS Tandon, 2012).

Table 2.20 Average crop responses to Zinc in field experiments

Crop	Trials averaged	Mean response to Zinc Kg/ha	Percent
Wheat	2447	420	18
Paddy	1652	540	23
Maize	280	470	16
Sorghum	83	360	41
Pearl Millet	236	190	13
Finger Millet	47	350	14
Chickpea	15	360	24
Black Gram	10	240	28
Groundnut	83	320	18
Soybean	12	360	11
Mustard	11	270	18
Potato	45	2.96 t/ha	17
Cotton	27	0.22	54

Source: HLS Tandon, 2012

Application of Zn to hybrid rice in the high yielding rice-wheat rotation increased paddy yield by 634 kg/ha and that of the following wheat by 286 kg/ha. These results averaged over 10 locations and 2 years show that annual grain production can be increased by 1 tonne/ha through zinc application (Tiwari, *et al.*, 2006).

In the case of potato, Zn application increased potato yield by 2.4-3.4 t/ha under different soil – climatic conditions (Grewal and Trehan, 1990). In horticultural crops, zinc application @200g zinc sulphate/ tree increased papaya yield by 47% while foliar spray with 0.5% zinc sulphate increased papaya yield by 12.5% (Ganeshamurthy et al., 2014). Combined spray of Zn and B (0.1%B) once in 60 days was most effective resulting in an increase in the number of fruits/tree by 31% as compared to 12.5% by Zn and 6% by B when applied separately.

EFFECT OF IRON ON CROP YIELD

Iron plays a significant role in various physiological and biochemical pathways in plants. It serves as a component of many vital enzymes such as cytochromes of the electron transport chain, and it is thus required for a wide range of biological functions. In plants, iron is involved in the synthesis of chlorophyll, and it is essential for the maintenance of chloroplast structure and function.

Crop responses to iron under field conditions have been reported for a number of crops (Malewar and Ismail, 1995, Takkar and Nayyar, 1984, Grewal and Trehan, 1990). This is because of the vast areas under alkaline-calcareous soils which are generally low in available Fe. Responses are generally low in the range of 300-600 kg/ha based on multi-location trials but would depend on the crop, available Fe status of soils and the yield potential.

Table 2.21 Average crop responses to Iron in field experiments

Crop	Response to Fe (kg/ha)	Crop	Response to Fe (kg/ha)	Crop	Response to Fe (kg/ha)
Wheat	780	Finger Millet	300	Cowpea	470
Rice	1880	Sorghum	580	Groundnut	230
Maize	450	Chickpea	450	Sunflower	470
Barley	450	Lentil	670	Potato	2759

Source: HLS Tandon, 2012

EFFECT OF BORON ON CROP YIELD

Boron (B) is a micronutrient critical to the growth and health of all crops. It is a component of plant cell walls and reproductive structures.

Boron plays a key role in a diverse range of plant functions including cell wall formation and stability, maintenance of structural and functional integrity of biological membranes, movement of sugar or energy into growing parts of plants, and pollination and seed set. Adequate B is also required for effective nitrogen fixation and nodulation in legume crops.

Information on the impact of Boron application on crop yields has grown over the years, most likely due to increasing incidence of boron deficiencies in soils (Tandon, 2009). Much of the information available is from Bihar, West Bengal and North-eastern region. Based on the data gathered from 58 trials, on an average, boron application increased the yield of rough rice by 310kg/ ha and that of wheat by 370 kg/ha which is the mean obtained from 36 trials. In Andhra Pradesh, B application increased the grain yield of Sorghum by 13% and that of maize by 20%. In Madhya Pradesh, application of 1kg/ha increased the yield of rainfed wheat by 900kg/ha (33%) (Rego et al., 2005). Subsequent compilations provided basically similar trends (Singh 2008). Crops for which results of 5 or more trials were documented are summarised in Table 2.22.

Table 2.22 Summary of Crop responses to Boron application

Crop	Trials	Average response to B over NPK	
		Kg/ha	Percent
Rice	107	320	16.6
Wheat	35	390	15.1
Maize	5	570	32.5
Chickpea	7	350	44.1
Groundnut	11	120	9.9
Sesame	5	90	23.9
Mustard	15	268	32.8

Source: HLS Tandon, 2012

EFFECT OF COPPER ON CROP YIELD

Copper is an essential metal for normal plant growth and development. Copper is required for many enzymatic activities in plants and for chlorophyll and seed production, and is an essential cofactor for many metalloproteins.

As compared to other micronutrients, there is much less information available on the effect of Cu on crop yields, probably due to much lower incidence of Cu deficiencies as compared to Zn or B. In 25 experiments from Bihar, crop responses to copper were obtained in 60% cases (Sakal et al., 1996). Rice responded most to foliar applied Cu in red and yellow, terai, medium black and sub- mountainous soils while wheat responded mostly in the laterites and mixed red black soil. Responses of potato to Cu were highest in hill soils, followed by black, alluvial, red and lateritic soils (Grewal and Trehan, 1990).

EFFECT OF MANGANESE ON CROP YIELD

Manganese (Mn) is an important micronutrient for plant growth and development and sustains metabolic roles within different plant cell compartments. It plays an important role in oxidation and reduction processes in plants, such as the electron transport in photosynthesis. Manganese also has played a role in chlorophyll production, and its presence is essential in Photosystem II. Manganese acts as an activating factor which causes the activation of more than 35 different enzymes due to the metabolic role of manganese in the nitrate-reducing enzyme activity and activation of enzymes which play roles in carbohydrate metabolism.

Average yield increases reported due to Mn application have been 360 kg/ha in rough rice, 560 kg/ha in wheat and 2026 kg/ha in potato (Grewal and Trehan, 1990, Takkar and Nayyar 1984). In the rice-wheat rotation where 20-30kg $MnSO_4$/ha was applied only to rice, led to an increase in rice yield by 590 kg/ha followed by a residual response of 136 kg/ha in terms of extra wheat yield. Out of the total annual yield increase, 81% was direct and 19% was residual. When similar treatment was applied to plots under the rice-rice rotation, the yield increase due to Mn application was 29 kg/ha in Kharif rice but 341 kg/ha in Rabi rice (Tiwari et al., 2006).

EFFECT OF MOLYBDENUM ON CROP YIELD

Molybdenum is a micronutrient that is directly involved in the metabolic functions of nitrogen in the plant. The transition metal molybdenum, in molybdate form, is essential for plants as a number of enzymes use it to catalyze most important reactions in the nitrogen acclimatization, synthesis of phytohormone, degradation of purine and the detoxification of the sulfite. There are more than known 50 different enzymes that need Mo, whether direct or indirect impacts on plant growth and development, primarily phytohormones and the N-metabolism involving processes.

Research on crop responses to Mo application is much less than for other micronutrients but significant yield increases due to Mo application have been reported in several cases. In field trials with potato, Mo application increased tuber yield by 1.2t/ha in alluvial soil, 1.3 t/ha in red and lateritic soil, 2 t/ha in hill soil and by 2.9 t/ha in black soil (Grewal and Trehan, 1990). Results from 63 trials with wheat showed a mean yield increase of 151 kg/ha (8.3%) to Mo application (Singh 2008). In acidic, high altitude soil in Andhra Pradesh, application of 1 kg sodium molybdate/ha increased maize grain yield by 1565kg / ha or 35%over the control. Responses of several crops to Mo application in acid soils have also been documented (Sarkar and Singh 2003). In Jharkhand, Mo application increased the yield of Lucerne by 9%, groundnut by 20%, soybean by 26% and of cauliflower by 33%. In Odisha, the yield of green gram increased by 40% while that of soybean in Karnataka increased by 26%due to Mo application. In Himachal Pradesh, application of 250gm sodium molybdate in cauliflower produced a yield increase of 21.4 t/ha, 60% more than in the control (HLS Tandon, 2012). Several reports of yield increase in oilseeds due to Mo application have been documented (Hegde and Sudhakar Babu, 2009). At Pantnagar, Mo application increased soybean yield by 332kg/ha (10.6% +) and 577 kg/ha (18.5%+) at an application rate of 0.5 kg/ha and 1 kg/ ha respectively.

Methods Of Supplementation

3.1 SPECIALTY FERTILISERS

India has achieved four fold enhancements in foodgrain production mainly due to increase in cropping intensity, growing of high-yielding varieties, shift in cropping pattern, increased use of high analysis fertiliser and irrigation. In green revolution and post green revolution era, crops were fertilised with N, NP, NPK and micronutrient application was almost ignored. This has caused mining of soil nutrients and resulted in depletion of native soil fertility. As a result, widespread deficiencies of nutrients like Zn, B, Fe and Mn were recorded as a major cause of stagnation in crop yield not only in cereals but in other crops like mustard, pearl millet, sugarcane and legumes also. If this trend of nutrition depletion continues, more and more areas would certainly suffer from secondary and micronutrient deficiencies.

Current soil fertility scenario of Indian soils indicates that 85-90 %, 70-80 %, 42 %, 49 %, 33% 15 %, 7 %, 6 % and 4 % soils of India are deficient in N, P, S, Zn, B, Fe, Mo, Mn and Cu, respectively (Muralidharudy *et. al.*, 2011 and Tandon, 2012). This scenario of nutrient deficiency is not uniform over the regions but location, region, soil and crop specific. Zn deficiency is mainly associated with the states of Maharastra (86%), Karnataka (72%) whereas, B deficiency is widespread in West Bengal (68%) and Bihar (38%). Other micronutrients such as Fe are mainly deficient in Karnataka (35%), Himachal Pradesh (27%). Fertility imbalance requires management of micronutrients deficiencies in view of decline in production of major crops is a cause of concern that requires immediate attention. Thus, there is need of specialty fertilisers for site and location specific remedy of nutrient deficiency (Gupta, *et. al.*, 2018).

The importance of speciality fertilisers in the context of Indian agriculture stems from the distorted nutrient ratio in the soils across different agro-climatic regions. Ideally, the NPK ratio of Indian soil should be 4:2:1 but as per the available records, it is around 6.7:3:1. The indiscriminate use of bulk fertilisers, especially the highly subsidized nitrogenous fertilisers, without considering the nutrient requirements and soil nutrient status has been reported in case of many crops. Further, nutrient use efficiency is also abysmally low in India more so under rainfed agriculture. Need of the hour is to improve the nutrient use efficiency, promote the use of balanced fertilisers with due importance to secondary and micro nutrients.

The fertiliser industry faces a continuing challenge to improve and increase the efficiency of their product use, particularly of nitrogenous fertilisers, and to minimize any possible adverse environmental impact. This can be achieved through improvement of fertilisers already in use, or through development of new specific fertiliser types. '*Specialty fertilisers are innovative sources of nutrients which are applied in special condition of soil and plant for special action in plant for achieving higher recovery, efficiency and economy*'. (Gupta, et. al., 2018)

TYPES OF SPECIALTY FERTILISERS

Understanding the necessity of maintaining the soil nutrient balance, the Government of India has formulated key policies and one among them is to promote the use of speciality fertilisers. The production of the speciality fertilisers in the country have hence, improved to the level of 0.1 lakh tonnes of fortified fertilisers, 0.3 lakh tonnes of water soluble fertilisers and 0.5 lakh tonnes of customised fertilisers as per 2015-16 statistics (Praveen and Aditya, 2017).

The various categories of specialty fertilisers are

1. Water Soluble Fertilisers
2. Slow and controlled release fertilisers
3. Fortified Fertilisers
4. Liquid Fertilisers
5. Stabilised Fertilisers
6. Customized fertilisers

3.1 a Water Soluble Fertilisers

Water soluble fertilisers is a category among speciality fertilisers meant for fertigation or nutrition of high value field crops and horticultural crops. These are, at present, one of the fastest growing agricultural inputs in the country. These fertilisers have varying ratio of primary, secondary and micronutrient with low salt index and are compatible with other agrochemicals (Gupta *et. al.*, 2018). They are used for both, fertigation as well as foliar application.

The FCO has created a separate category of "100% Water Soluble Fertilisers" wherein several NP/NPK grades with 100% solubility in water are described. In addition to these, there is an array of NPKS grades varying in nutrient ratios with 12- 18% S.

Popular high water soluble specialty fertilisers are:

1. Potassium Nitrate (13-0-45)
2. Mono Potassium Phosphate (0-52-34)
3. Potassium Magnesium Sulphate
4. Mono Ammonium Phosphate (12-61-0)
5. Urea Phosphate (17:44:0)
6. Calcium Nitrate
7. Sulphate of Potash

Table 3.1 Forms of N in NP / NPK complexes (100% water soluble)

Fertiliser	Total N (%)	Percentage of total N as		
		Ammoniacal	Nitrate	Amide
6-12-36	6	25	75	0
13-5-25	13	46	54	0
13-40-13	13	66	34	0
24-24-0	24	25	0	75
28-28-0	28	20	0	80
18-18-18	18	45	55	0
19-19-19	19	24	21	55
20-20-20	20	15	25	60

Consumption of WSF increased drastically from a meagre 1200 tonnes in the year 1995-96 to 1.7 lakh tonnes in the year 2015-16. Several grades of water soluble fertilisers are available among which calcium nitrate is the most popular one with a share of 25.4 per cent in sales of total water soluble fertilisers, followed by 19-19-19 (19.3%), potassium nitrate (12.6%), SOP (9.9%) and MAP (9.7%) (Praveen and Aditya, 2017).

The growth in hi-tech farming in the country and the adoption of better irrigation practices like drip and other micro irrigation methods, are the key drivers for increased consumption of these fertilisers in India. Improvement in the fertiliser use efficiency through the use of water soluble fertilisers in fertigation contributes significantly to the increase in crop yield. For example, the nitrogen use efficiency in fertigation is 95 per cent in comparison to 30-50 per cent in soil application. Similarly, the phosphorous and potassium use efficiencies are also higher in fertigation by about 25 and 30 per cent respectively in comparison to soil application. Maharashtra, Karnataka, Andhra Pradesh (undivided), Tamil Nadu and Kerala are the leading consumers of water soluble fertilisers.

3.1 b Slow And Controlled Release Fertilisers

Slowing the release of plant nutrients from fertilisers can be achieved by different methods and the resulting products are known as slow- or controlled-release fertilisers. Slow- and controlled-release fertilisers may contain only N or K, NP or NK (with different forms of K), NPK or NPK plus secondary nutrients and/or different micronutrients (Shoji and Gandeza, 1992).

Fig. 3.1a: Slow and Controlled Release fertiliser

With controlled-release fertilisers, the principal method is to cover a conventional soluble fertiliser with a protective coating (encapsulation) of a water-insoluble, semipermeable or impermeable-with-pores material. This controls water penetration and thus the rate of dissolution, and ideally synchronizes nutrient release with the plants' needs.

Controlled-release fertilisers improve the uptake of nutrients by plants through synchronized nutrient release, and significantly reduce possible losses of nutrients, particularly of Nitrate-N by leaching and volatilization losses of ammonia. This substantially decreases the risk of environmental pollution (Koshino, 1993; Mikkelsen *et. al.*, 1994; Rietze and Seidel, 1994; Shaviv, 1996, 2005; Wang, 1996; Zhang *et. al.*, 2001; Shoji, 2005; Ma *et. al.*, 2007; Zhang, 2007). A reasonably good prediction of nutrient release is possible with controlled-release fertilisers coated with hydrophobic materials, particularly polymer-coated fertilisers because they are less sensitive to soil and climatic conditions (Shaviv, 1996, 2005; Shoji, 1999, 2005).

Some slow and controlled release fertilisers mentioned in the FCO are

Urea Formaldehyde:

Urea formaldehyde is the organic N compound mainly used for the slow release of nitrogen.

Based on the release of nitrogen from these Urea formaldehyde products, the compounds are again grouped into three fractions:

Cold water-soluble N—it contains mainly urea dimers, soluble short Urea formaldehyde chains. In this fraction, N is readily available.

Hot water-soluble N—it contains methylene urea and chains of intermediate length. N is slowly released into the soil.

Hot water-insoluble N—it contains intermediates and long chain and extremely slow decomposing or practically unavailable nitrogen.

The decomposition of UF is mainly due to the microbial action. So the release of nitrogen from these compounds mainly depends on soil properties such as biological activity, pH, moisture content, and temperature

Sulphur coated urea (SCU):

SCU was developed in 1961 by giving each prill of urea a physical coating of elemental sulphur. It is regarded as a slow release source of nitrogen meant to release nitrogen only after the disintegration of the sulphur coating. Its efficacy as a slow release fertiliser depends upon the uniformity and thickness of the S coating.

Sulphur coated NPK complex:

NPK fertilisers coated with S have been developed wherein molten sulphur is sprayed on the granules of the fertilisers to be coated. Then the surface of the sulphur layer is coated with wax to fill the pores in the coating. Finally, a mineral coating is given. Eight types of S coated NK and NPK fertilisers with varying nutrient ratios for different crops have been described (Miyata, 1989).

Neem coated urea (NCU):

NCU has been developed and is being produced on a commercial scale by several urea manufacturers in India. Oil extracted from the neem seeds is used to coat urea during manufacture. The granular product contains 46%N as the weight of the coating material is negligible.

There are also conventional coated/ encapsulated soluble fertiliser materials with rapidly plant-available nutrients, which after granulation, prilling or crystallization are given a protective, water-insoluble coating to control water penetration and thus dissolution rate, nutrient release and duration of release.

There are three main groups of coated/encapsulated fertilisers, based on the following coating materials:

- ☐ sulphur,
- ☐ sulphur plus polymers, including wax polymeric materials, and
- ☐ polymeric/polyolefin materials.

Coated/encapsulated fertilisers offer flexibility in determining the nutrient release pattern (Fujita *et. al.*, 1983; Shoji and Takahashi, 1999). They also permit the controlled release of nutrients other than nitrogen. Nyborg *et. al.* (1995) found in greenhouse and field tests that slowing the release of fertiliser P into the soil by coating fertiliser granules (polymer coating) can markedly increase P recovery by the crop in the year of application and improve yield.

Polymer coatings may either be semi-permeable or impermeable membranes with tiny pores. The main problems in the production of polymer-coated fertilisers are the choice of the coating material and the process used to apply it (Fujita and Shoji, 1999; Goertz, 1993; Hahndel, 1986; Moore, 1993; Pursell, 1992, 1994, 1995). The nutrient release through a polymer membrane is not significantly affected by soil properties, such as pH, salinity, texture, microbial activity, redox-potential, ionic strength of the soil solution, but rather by temperature and moisture permeability of the polymer coating. Thus, it is possible to predict the nutrient release from polymer-coated fertilisers for a given period of time much more reliably than, for instance, from SCU (Fujita and Shoji, 1999; Shaviv, 2005; Shoji and Gandeza, 1992).

The reasonable good prediction of long-term nutrient release from some types of controlled-release fertilisers makes it possible to develop software progammes for their use on different crops and for various soil and growing conditions. Such software programmes can be very reliable for polymer-coated fertilisers, because, there is a reasonable good correlation between temperature, release of nutrients and plant growth (Shoji, 2005). In intensive vegetable production, slow- or controlled-release fertilisers offer onetime application of the fertiliser with multiple cropping, e.g. multi-cropping of spinach, lettuce, Chinese cabbage, broad beans, broccoli, etc. in Japan. They also give the possibility to enhance the quality and safety of vegetables and farm produce: e.g. low protein in rice, high protein in wheat, high sugar and ascorbic acid with low nitrate and oxalic acid in leafy vegetables (Shoji, 2005)

Fig. 3.1b: Slow and Controlled Release fertiliser

3.1 c Fortified Fertilisers

Fortified fertilisers are combination of multi-nutrient carriers which satisfies the crop's nutritional demand based on area, soil and growth stage of a plant. A fortified fertiliser strengthens the nutrient supply to the plant and provides site specific nutrient management for achieving maximum fertiliser use efficiency for the applied nutrient in a cost effective manner (Gupta *et. al.*, 2018). There is growing interest in fortified fertilisers. This is primarily due to the increasing incidence of deficiencies of sulphur and several micronutrients and the fact that fortification facilitates the uniform application of small amounts of micronutrients without going in for separate application.

At present, 23.1 lakh tonnes of neem coated urea is produced in the country and 22.9 lakh tonnes are sold. Several other fortified/coated fertilisers are also available in the market with different nutrient compositions. Zincated urea, zincated phosphate, NPK complex fertiliser fortified with zinc and boron, bentonite sulphur with zinc, DAP fortified with zinc, DAP with boron, calcium nitrate with boron, SSP with zinc etc. being the major ones (Praveen and Aditya, 2017).

3.1 d Liquid Fertilisers

Speciality liquid fertilisers are high analysis, totally water soluble fertilisers available in mono, double and multi nutrient combinations that can be applied to plants through fertigation or foliar application to maximize fertiliser use efficiency and crop productivity, minimize production cost and to improve quality of crop and its produce. Some of the liquid fertilisers mentioned in the FCO are Ammonium polyphosphate (10-34-0), Urea Ammonium Nitrate (32%N) and Ammonium Thiosulphate (12% N and 26% S).

Some of the other commonly available liquid fertilisers are:

1. Aqua-ammonia
2. Vermiwash
3. Sea weed extract
4. Potassium thiosulphate solution
5. Boron Ethanolamine 6.5% & Boron (B): 15.0%

3.1 e Stabilized Fertilisers

Urease inhibitors and nitrification inhibitors are commonly known as stabilized fertilisers. Urease inhibitors prevent or suppress the transformation of Amide-N in urea to ammonium hydroxide and ammonium through the hydrolytic action of the enzyme urease over a certain period of time.

By slowing down the rate at which urea is hydrolysed in the soil, volatilization losses of ammonia to the air (as well as further leaching losses of nitrate) is either reduced or avoided. Thus, the efficiency of urea and of N fertilisers containing urea (*e.g.* urea ammonium nitrate solution), is increased and any adverse environmental impact from their use is decreased. N-(n-Butyl) thiophosphoric triamide (NBPT), phenyl phosphorodiamidate (PPD/ PPDA) and hydroquinone are probably the most thoroughly promising urease inhibitors (Kiss and Simihaian, 2002).

Nitrification inhibitors delay the bacterial oxidation of the ammonium ion (NH_4^+) by depressing over a certain period of time (four to ten weeks) the activity of *Nitrosomonas* bacteria in the soil. These bacteria transform ammonium ions into nitrite (NO_2^-), which is further transformed into nitrate (NO_3^-) by *Nitrobacter* and *Nitrosolobus* bacteria. The objective of using nitrification inhibitors is to control the loss of nitrate by leaching or the production of nitrous oxide (N_2O) by denitrification from the topsoil by keeping N in the ammonium form longer and thus increasing N-use efficiency. Most commonly used nitrification inhibitors are 1-chloro-6-(trichloromethylpyridine)- Nitrapyrin, Dicyandiamide (DCD), Thiourea, 2-amino-4-chloro-6-methyl-pyramidine (AM) and Neem (Gupta et.al., 2018).

3.1 f Customized Fertilisers

According to FCO, Customized fertilization means the use of the fertilisers best management practices (BMPs) and is generally assumed to maximize crop yields while minimizing unwanted impacts on the environment and human health. Present scenario of nutrient deficiency of our country is not uniform over the regions but location, region, soil and crop specific. Thus, it requires the soil, crop and action specific fertilisers best management practices (BMPs).

Customized fertilisers are such specialty fertilisers which are being made for specific condition of soil and recommended for special crops and hence in this way promote site specific nutrient management. These are soil-crop-climate based fertiliser and are less influenced by soil, plant and climatic conditions that lead to more uptake of nutrients and less loss of nutrient maximising fertiliser use efficiency in a cost effective manner. Scientific principles such as geo-referencing, sampling of soil, plant and water sample of the chosen area, defining management zones, yield targeting, calculating nutrient requirement, blending of nutrients based on the generated information are used as an ultimate guiding factor in deciding the grades of customized fertilisers. Total sales of customised fertilisers showed a gradual increase from 0.19 lakh tonnes in the year 2008-09 to 0.45 lakh tonnes in 2015-16 (Praveen and Aditya, 2017)

CUSTOMIZED FERTILISER FORMULATIONS AVAILABLE IN INDIA

There are about 36 formulations approved by fertiliser control order of India. Some of the popular customized fertilisers are as mentioned in Table 3.2:

Table 3.2 Customised fertiliser formulations available in India (Gupta et al., 2018)

Sr. No	Grade	Recommended Crop	Recommended Area
1	15:32:8:0.5 (Zn)	Paddy	East and West Godavari of AP
2	18:27:14:05 (Zn)	Maize	East and West Godavari of AP
3	12:24:0:0.5 (Ca)	Cotton	Karimnagar, Warangal & Nizamabad
4	8:16:24:6:0.5:0:1.5	Potato	Western and northern districts of Uttar Pradesh
5	7:20:18:6:0.5	Sugarcane	Western and northern districts of Uttar Pradesh
6	12:26:18:0.5:5	Wheat and Paddy	Pratapgarh, Barabanki, Faizabad, Jaunpur and Raebareilli

Need for Specialty Nutrition

INDIAN AGRICULTURE *is laden with paradox- the most striking being the paradox of productivity. We have one of the largest cultivable land masses on Earth. Although productivity gains were sustained in the 1990s after the liberalization process began, the yield rates for most of the crops in India are far below comparable rates in a number of other countries.*

One of the major reasons for the poor yields per unit of land area is the inefficiency in the use of fertilisers and poor cultivation practices leading to imbalanced nutrition. The use of modern farming practices on a wider scale and integrated nutrient management practices are essential if India's farmers wish to produce crops in line with the observed global standards of quantity and quality. A consequent increase in production efficiency, upon following the best practices in crop nutrition, would enhance value added activities in agriculture through agro processing and exports of agriculture and agro based products. These activities in turn would increase income and employment in the industrial processing sector. Thus globalizing agriculture has the potential to transform subsistence agriculture to commercialized agriculture and to improve the living conditions of the rural community.

When addressing the need for balanced fertilization using specialty farming techniques, it must be borne in mind that the potential for increasing productivity of land already under cultivation is far greater than that for bringing more land under cultivation. The inherent limitation in this approach is that plant nutrient deficiencies in both irrigated and rain-fed farms will increase. The problem of eliminating these deficiencies is already a major problem today and it will aggravate in the future.

Integrated nutrient management has spinoffs as well. The most crucial is the improvement in the plant physiology that builds levels of resistance and reduces the incidence of disease and pest attacks. Increasing resistance through efficient nutrition programmes can thus reduce the application of harmful (and expensive) pesticides and make farming more productive, sustainable and environment friendly.

Nutrient Requirement of crops

The application of nutrients like nitrogen (through urea), phosphorus (through DAP) and potassium (through SOP) is common place. These are nutrients required in large quantities and they are well recognized as major nutrients. In fact, there are several instances observed of their rampant overuse, which is a serious cause for concern. A major problem is faced with urea. Nitrogen in urea has an immediate greening effect on the crops and the farmers tend to blindly apply urea as soon as symptoms of chlorosis appear. The field becomes green overnight and the farmer feels that he has effectively managed the problem at hand. However, the green fades in a matter of a few days. This "boot-polish" effect of urea has led to its unscientific usage and has upset the nutritional balance, as the excess nutrients applied tend to reduce the efficiency of uptake of the other nutrients present in the soil, compounding the problem of deficiencies. However, N, P and K are not the only nutrients required by plants. Plants need several inorganic elements for synthesis of their own cellular compounds for seed/grain formation, maturity, formation of enzyme systems and as energy sources for their growth. These elements are essential nutrients having specific functions. Neither can a plant complete a healthy life cycle in the absence of even one of the essential nutrients, nor can it be replaced.

Calcium, magnesium and sulphur are considered secondary nutrients and several other nutrients called micronutrients though required in smaller quantities, are equally important for good growth and development of plants. The essential micronutrients are boron (B), chlorine (Cl), copper (Cu), iron (Fe), manganese (Mn), molybdenum (Mo) and zinc (Zn). Plants take up all these nutrients simultaneously and their requirements vary with the type of plants, growth stages, yield potential and the like.

A recognition of the specific role that each of the secondary and micronutrient plays in the growth and development of plants is crucial. Moreover, widespread awareness needs to be developed throughout the Nation, using the private and government extension machinery. Most micronutrients, especially the transition metals Zn, Fe, Cu, Mn and Mo are constituents of many metal enzymes and they function in key metabolic events such as chlorophyll synthesis, photosynthesis, respiration, protein synthesis, nitrogen fixation, assimilation of nitrates and sulphate and the like.

Micronutrients are therefore responsible for key processes of plants including efficient utilization of even major nutrients.

A wide gap in Consumption

Awareness of the need and utility of secondary and micronutrients is still in its infancy in India, though companies have been promoting the concept since 1975. Their non-use or under-use increases the imbalances in plant nutrition and serves as a severe limiting factor to enhance farm productivity in India. This imbalance needs to be overcome immediately if farm productivity levels need to be enhanced.

The use of traditional, age-old crop nutrition and crop management practices needs urgent change. Nations where higher yields have been observed have uniformly seen farmers adopting new generation farming techniques, including drip- irrigation, hydroponics, protected cultivation, use of water soluble complex specialty fertilisers, genetically modified seeds and the like. Farmers using these new technologies have, to a certain extent, gained control over the environment and consequently, reduced the risks involved in agriculture.

Formulating a balanced crop nutrition programme

In the process of formulating an ideal balanced crop nutrition programme, it is essential for the farmer to begin with a thorough soil health study. Progressive plant nutrition companies have equipped their laboratories to assist farmers in getting an accurate soil profile with an accurate assessment of nutrient deficiencies. State Government soil testing laboratories are also providing similar facilities to farmers at nominal charges. The next step is to identify and evaluate the various options available to supply nutrition to the crops.

a) Use of water solubles

The use of traditional granular NPK fertilisers (urea, DAP, etc.) is declining in popularity in all progressive nations of the world. Progressive farmers are shifting to water-soluble complex NPK fertilisers which are a cost-effective option.

In terms of nutrient absorption, foliar fertilization is seen to be from 8 to 20 times as efficient as ground application. Foliar feeding is possible very effectively using water soluble complexes. Moreover, the numbers of formulations available are diverse with different combinations of nutrients, thus allowing a farmer to custom design the nutrition programme. This is unlike the very limited combinations available with traditional, granular NPK fertilisers. These complex fertilisers are highly pure (97-99%), completely water soluble and efficiently absorbed by the crops making them ideal for foliar fertilization programmes at very low doses. Also, where farmers are using micro irrigation systems, the only feasible option available is to use such specialty water soluble fertilisers.

As far as the micronutrients are concerned, traditional usage of inorganic forms of micronutrients has its inherent disadvantages. These include the formation of insoluble salts and acids in the soil after interacting with air and soil components. Inorganic nutrients are vulnerable to leaching and fixation rendering them unavailable for absorption by the crops. Studies indicate that 60% of nutrients are wasted if applied as inorganic sulphates.

b) Chelated micronutrients

The usage of chelated forms of micronutrients is the most effective, environment friendly option to farmers worldwide. A chelated micronutrient has the metallic ion of the nutrient (Zn, Fe, Mn, Cu, etc.) surrounded by a carefully designed protective cage structure. This creates an inert organo-metallic complex ensuring that the nutrient can be delivered unharmed and unchanged (by various interactions) at its destination – the crop. Thus, the specially engineered bond formed between the nutrient and the chelating agent locks in the nutrient ions and provides optimum agronomic use of the micronutrient.

High quality chelates are in use in certain parts of India and have shown extremely promising results. The low dosages required leads to a reduction in the cost of application per unit of land area. 1 kilo of a chelated micronutrient is seen to give better yields compared to 20 kilos of an inorganic form of the same micronutrient.

c) Tools and techniques available

The tools and techniques to ensure the improvement of farm productivity are now available in India. Companies are investing a great deal on Research and Development to ensure the availability of state of the art products and farming systems. The Universities and Research Institutions have tried tested and documented the immense benefits from the usage of such technology. Such research is continuing and draws on the favourable experiences of countries around the world. It is now time to spread awareness. The farmers of India have a right to know that there are cost effective specialty plant nutrition solutions available which have the potential to significantly increase the productivity of their farms. They have a right to be educated in the efficient usage of such technologies. In return, India can be assured of a renewed ever-green productivity revolution in Her agriculture sector.

Dr. Rahul Mirchandani
CMD, Aries Agro Ltd,
IMMA Souvenir, 2020

3.2 CHELATION AND CHELATED NUTRIENTS

Micronutrients are essential nutrients required by plants in small quantities. They are as important to plant nutrition as primary and secondary macronutrients. A lack of any one of the micronutrients in the soil can limit growth, even when all other nutrients are present in adequate amounts.

The three main sources of micronutrients are inorganic sources, synthetic chelates and organic complexes. Inorganic sources are most commonly used due to their ready availability and water solubility. If these elements are added to fertiliser or soil on their own they will often become bound up in the soil and will not be available to the plants. Hence, it has become imperative in today's times to develop innovative, environmentally benign technologies to protect human health and ecosystems. Chelates prevent this from occurring and help plants absorb more nutrients.

The word chelate is derived from the Greek word for "claw". The word chelate was first used by researchers in the 1920s because it describes the principle of grasping and holding something, which is essentially what occurs in the process of chelation. In fertiliser technology, it refers to inorganic nutrients that are enclosed by an organic molecule. A chelate is a specific kind of chemical compound that is more easily dissolved and absorbed than other types of molecules and chemical compounds. It is made up of a metal ion and a chelating agent that creates multiple soluble bonds to the ion. This chemical structure makes it easier for chelates to be absorbed in a solution. Metals bound in chelate rings have essentially lost their cationic characteristics and are less prone to precipitation in some chemical reactions.

ROLE OF CHELATES IN NUTRIENT AVAILABILITY

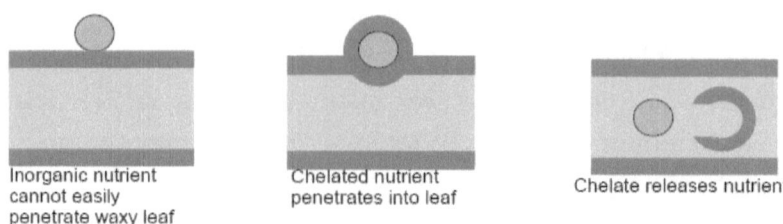

Fig. 3.2: Role of chelates in nutrient availability

Many trace elements in their basic form(s) are unavailable to plants. This is basically due to the fact that these metals are positively charged (Fe, Mn, Zn, Cu, etc.). The pores (openings) on the plants' leaves and roots are negatively charged. As a result, the element can't enter the plant due to the difference in charges. However, if a chelate is added with an element like iron it effectively encapsulates (surrounds) the metal/mineral ion and changes the charge into a negative or neutral charge, allowing the element to enter the pore and travel into the plant. Some chelating agents may only have the ability to partially surround an element and should be referred to as a "complexes", while those that completely surround the mineral are true chelates. The bond between the organic chemical and the inorganic nutrient must be strong enough to protect the nutrient, but must be weak enough to release the nutrient once it gets into the plant. Also, the chelating agent must not be harmful to plants. Not all nutrients can be chelated. Iron, zinc, copper, manganese, calcium and magnesium can be chelated, the other nutrients cannot.

Advantages of Chelates Over Tradition Forms

- Lower quantities of chelates are required as they are easily assimiable
- Chelates are cost effective
- Chelates are more easily absorbed by plant roots or leaves
- Chelates are easily translocated within the plant.

- Under alkaline conditions, chelated iron, zinc, manganese and copper is a better way to provide micronutrients to a crop.
- Chelates are compatible with a wide variety of pesticides and liquid fertilisers, as chelates do not react with their components. Most chelates can be mixed with dry mixes and liquid fertilisers.
- Chelates are not readily leached from the soil as they adsorb on to the surface of soil particles

SYNTHETIC CHELATING AGENTS

Several organic substances (chelating agents) are used to produce chelates. Examples of synthetic chelating agents include ethylene diamine-tetra-acetic acid (EDTA), diethylene-triamine penta-acetic acid (DTPA), ethylene-diamine-di- (o-hydroxyphenylacetic acid) (EDDHA). EDTA is the most common synthetic chelating agent and is used for both soil and foliar applied nutrients. EDTA has four points of connection to the elements it chelates. Different chelates have varying numbers of points of connection. EDTA is better suited to slightly lower than neutral ph levels. Iron often becomes deficient at higher ph values such as those typically associated with rockwool or mineral soils.

Diethylenetriaminepentaacetate (DTPA) is a chelating agent better suited to high ph levels. As the chemical name suggests, it has five (penta, i.e. Pentagram) points of connection to the element it chelates. It is more effective than EDTA but is usually more expensive.

Several studies suggest that ethylenediaminedihydroxy-phenylaceticacid (EDDHA) is a superior synthetic chelating agent. Its relatively high cost prohibits it from being added to many synthetic fertiliser formulations.

NATURALLY OCCURRING CHELATING AGENTS

Some naturally occurring chelating agents include products of organic matter decomposition like humic acids, organic acids, amino acids, sugar acids and derivatives, phenols, polyflavonoids, siderophores and phyto siderophores. One of the greatest advantages in adding a biological chelate is that unlike the synthetic chelates, the organic chelates can be absorbed into the plant.

Fulvic acid is a biological chelating agent. Fulvic acids are lighter in weight and more biologically active than their precursor, humic acids.

Another category of biological chelating agents includes amino acids. Amino acids can function very well in chelating, as they are similar to a magnet in that they have both negative and positive charge (like north and south poles). In chelation, amino acids form a five point bond with the mineral element. This allows amino acid chelates to function well due to their relative stability. When the amino acid chelates reach the cell membrane, they are recognised by the mechanisms of absorption as a source of organic N. As a result, entire amino acid chelate is taken into the cell very rapidly and efficiently. However, because of its efficient absorption, much lower doses of true amino acid chelates need to be applied to the crops. True amino acid chelates are available as Zn, Fe, Mn, Cu and B complexed forms

Balanced Crop Nutrition Using Metal Chelates

What is balanced nutrition

Balanced fertiliser use refers to application of essential plant nutrients in optimum quantities and in right proportional through appropriate method and time of application suited for specific crop and agronomic situation. Balanced fertiliser rates of application differ from area to area and crop to crop. Soil testing is one of the most important tools to practise balanced fertilization.

Earnst Scenario: *Soil test summarizes that Indian soils have low to medium available P and 60% medium K whereas N continues to be universally deficient. 47 percent soils are efficient in Zn, 12% Cu, 4% Mn. In recent years' phenomenal increase in S deficiency has been witnessed specially in intensive cropping system where fertilisers devoid of S are used.*

Thus in situation where besides NPK the nutrients such as Zn, Fe, Mn, Cu and S are also becoming limiting factors. It is unthinkable to have a sustained food security without balanced and integrated use of nutrients. A balanced and judicious use of fertilisers is a key to efficient nutrient use and for maintaining soil productivity.

How to know when a plant needs nutrients?

Soil Analysis, Tissue Analysis, reduced vegetative growth – dwarf plants, Leaves become yellow or get curled, Increased Pest Attack, Yield drop Y-o-Y basis, Plant mortality rate is high after transplanting and absence of earth-worms in soil.

Why Micro Nutrients?

Out of 17 nutrients established as essential for plant growth, 8 are required in small quantities and therefore called micronutrients. They are Zinc, Boron, Iron, Manganese, Molybdenum, Nickel and Copper.

Micronutrients are needed in very small amounts, but are still essential for growth of plants. Micronutrient deficiencies vary with regions based upon soil, mineralogy and climate and often can be corrected by adjusting pH.

The importance of fertilization of crops with micro-nutrients is increasing mainly because of greater removal from the soil, intensive liming of soil, intensive drainage of soil, higher use of nitrogenous, Phosphatic and potassic fertilisers etc.

Fertiliser- Micronutrient Use Efficiency (MUE)

Micronutrient use efficiency by crops can be defined as the yield of biomass per unit input of fertiliser or nutrient content (Hawkesford et. al., 2014). Alternatively, MUE can be defined as the relative proportion of fertiliser-MN added to the soil or leaves that is absorbed by crops. In this we define MUE as the amount of the added fertiliser - MN that ends up in the crop, as biomass yield may be influenced by other factors, such as plant genetics and weather. The MUE also associated with the transport, usage and storage within the plant and indirectly with MN fate in the environment.

Iron, Mn, Zn and Cu rapidly react in soil components via oxidation and / or precipitation. Applying these nutrients directly to the soil is inefficient because in soil solution they are present as positively charged metal ions and will readily react with oxygen and/ or negatively charged hydroxide ions. Accordingly, the MUE of inorganic MN fertilisers ranges from 2.5% to 5% (singh2008; Samra, 2006). Higher rates of addition are required when MN are broadcasted relative to band application (Kaiser, 2011).

An important practical issue associated with MUE is uneven spatial distribution of the elements. Thus far chelated fertilisers appear to help improve MUE (Liu et. al., 2012). Low MUE increases crop production costs for farmers and adversely influences their content in crops, in addition to its negative effect on environmental health.

How to increase MUE?

There are several methods to improve MUE which include selection of micronutrient sources, Chelates, Foliar application, considering nutrient interactions & soil pH, Mixed fertiliser, Controlled release of micronutrient, microencapsulation and low solubility products and Nano - encapsulated micronutrients.

There are three main classes of micronutrient fertilisers: inorganic, Synthetic chelates and natural organic complexes. Inorganic sources consist of oxides, carbonates and metallic salts such as sulphates, chlorides and nitrates. Sulphates are the most common metallic salts used in the fertiliser industry because of their high water solubility and plant availability. Less soluble oxides must be ground finely or partially acidulated with sulphuric acid to form oxy-sulfates in order to increase their effectiveness. Metal-ammonia complexes such as ammoniated Zn sulphate decompose readily in soils and provide good agronomic effectiveness.

What is Chelation?

Think of the peel of an orange. The peel protects the fruit from dirt, bugs, or drying out until you are ready to consume it. This is similar to how a chelate works. A chelate occurs when two chelating agents attach to an ion. Like the orange peel, the chelate protects the nutrient from combining with other elements or losing nutrient value for absorption.

Chelates provide many benefits including increased nutrient mobility, nutrient precipitation prevention, the avoidance of nutrient leaching, and more. We will explore these benefits along with how chelates help plant and animal nutrition but first we need to understand the science of chelation.

Plant nutrients are one of the environmental factors essential for crop growth and development.

Nutrient management is crucial for optimal productivity in commercial crop production. Out of 17 nutrients established as essential for plant growth 8 are required in small quantities and those nutrients are in concentrations of ≤ 100 parts per million (ppm) in plant tissues which are described as micronutrients they include iron (Fe), zinc (Zn), manganese (Mn), copper (Cu), boron (B), chlorine (Cl), molybdenum (Mo), and nickel (Ni). Micronutrients such as Fe, Mn, Zn, and Cu are easily oxidized or precipitated in soil, and their utilization is, therefore, not efficient. Chelated fertilisers have been developed to increase micronutrient utilization efficiency.

What is chelated fertiliser?

The word chelate is derived from the Greek word chelé, which refers to a lobster's claw. Hence, chelate refers to the pincer-like way in which a metal nutrient ion is encircled by the larger organic molecule (the claw), usually called a ligand or chelator. Each of the listed ligands, when combined with a micronutrient, can form a chelated fertiliser. Chelated micronutrients are protected from oxidation, precipitation, and immobilization in certain conditions because the organic molecule (the ligand) can combine and form a ring encircling the micronutrient. The pincer-like way the micronutrient is bonded to the ligand changes the micronutrient's surface property and favours the uptake efficiency of foliarly applied micronutrients.

Why is chelated fertiliser needed?

Applying nutrients such as Fe, Mn, Zn, and Cu directly to the soil is inefficient because in soil solution they are present as positively charged metal ions and will readily react with oxygen and/or negatively charged hydroxide ions (OH^-). If they react with oxygen or hydroxide ions, they form new compounds that are not bioavailable to plants. Both oxygen and hydroxide ions are abundant in soil and soilless growth media. The ligand can protect the micronutrient from oxidization or precipitation. Because soil is heterogeneous and complex, traditional micronutrients are readily oxidized or precipitated. Chelation keeps a micronutrient from undesirable reactions *in solution and soil. The chelated fertiliser improves the bioavailability of micronutrients such as Fe, Cu, Mn, and Zn, and in turn contributes to the productivity and profitability of commercial crop production.*

There are many naturally occurring chelating agents that are products of organic matter decomposition such as organic acids, amino acids, lignosulfonates, ligninipolycarboxylates, sugar acids and derivatives, phenols, polyflavonoids, siderophores and phyto siderophores. Many chelating agents have been developed synthetically. Both classes of chelating/ complexing agents increase micronutrient solubility. In the soil, plant roots can release exudates that contain natural chelates. The nonprotein amino acid, mugineic acid, is one such natural chelate called phytosiderophore (phyto: plant; siderophore: iron carrier) produced by graminaceous (grassy) plants grown in low iron stress conditions. The exuded chelate works as a vehicle, helping plants absorb nutrients in the root-solution-soil system. A plant excreted chelate forms a metal complex (i.e., a coordination compound) with a micronutrient ion in soil solution and approaches a root hair. In turn, the chelated micronutrient near the root hair releases the nutrient to the root hair. The chelate is then free and becomes ready to complex with another micronutrient ion in the adjacent soil solution, restarting the cycle.

Which crops often need chelated fertilisers?

Vegetable and fruit crop susceptibility to micronutrients differs significantly. Soil pH is a major factor influencing micronutrient bioavailability; therefore, if soil pH is greater than 6.5, then the soil may have limited micronutrient bioavailability and chelated fertilisers may be needed.

Soil pH and chelated fertiliser requirements in commercial crop production

Soil pH < 5.3	Soil pH ranges from 5.3 to 6.5	Soil pH > 6.5
No chelated fertilisers are needed.	Chelated fertilisers may be needed	Chelated fertilisers are needed

At soil pH 5.3 or lower, soil can generally provide sufficient Fe, Cu, Mn, and Zn. In the soil pH range from 5.3 to 6.5, highly susceptible crop species may need chelated fertilisers. At soil pH 6.5 or greater, most crops need chelated fertilisers.

Which chelated fertiliser should be used?

Each of the ligands can form a chelated fertiliser with one or more micronutrients. The effectiveness and efficiency of a particular chelated fertiliser depends on the pH of the plant growth medium.

Common synthetic and natural chelate compounds (ligands) (Havelin et al., 2005, Sekhon, 2003)

Abbreviation	Name	Formula
CDTA	Cyclohexanediaminepentaacetic Acid	C14H22O8N2
CIT	Citric Acid	C6H8O7
DTPA	Diethylenetriaminepentaacetic Acid	C14H23O10N3
EDDHA	Ethylene diamine-N,N'-bis(2-hydroxyphenylacetic Acid	$C_{18}H_{20}O_6N_2$
EDTA	Ethylenediaminetetraacetic acid	$C_{10}H_{16}O_8N_2$
EGTA	Ethylene glycol-bis(β-aminoethyl ether)-N,N,N',N'-tetraacetic Acid	$C_{14}H_{24}O_{10}N_2$
HEDTA	Hydroxyethylethylenediaminetriacetic Acid	C10H18O7N2
NTA	Nitrilotriacetic Acid	$C_6H_9O_6N$
OX	Oxalic Acid	C2H2O4
PPA	Pyrophosphoric Acid	H4P2O2
TPA	Triphosphoric Acid	H5P3O10

Chelated fertilisers, formula, and nutrient content (%)

Source	Formula	Nutrient (w/w, %)
Iron Chelates	NaFeEDTA	5-14
	NaFeEDDHA	6
	NaFeDTPA	10
Copper Chelates	Na2CuEDTA	13 Cu
	Na2CuHEDTA	9
Manganese Chelates	Na2MnEDTA	5-12 Mn
Zinc Chelates	Na2ZnEDTA	14 Zn
	Na2ZnHEDTA	9-13 Zn
Natural Organic Materials		5-10 Fe, 0.5 Cu, 0.2 Mn, 1-5 Zn

Advantages of Chelates over Tradition Forms

The chelated forms of micro nutrients have a number of advantages over more traditional forms of trace elements such as oxides and sulphates:

1. Much lower quantities are necessary compared to inorganic compounds because they are completely assimiable by crops. Chelates are thus cost effective even though they are a little more expensive.
2. Chelates are much more easily absorbed by plant roots or leaves because chelates are of organic nature. The chelation process removes the positive charge from the micro nutrients allowing the neutral or slightly negatively charged chelates to slide through the pores on the leaf and root surface more rapidly. Since these pores are negatively charged, positively charged micro nutrients would normally be 'fixed' at the pore entrance would be difficult to be assimilated by plants. When neutral chelated micronutrients are used there would be no such restriction barriers.
3. Chelates are more easily translocated within the plant as their action is partly systemic.
4. Chelates are easily assimilated within the plant system.
5. The chances of 'scorching' of crops while using chelates is less because they are organic substances.
6. Under alkaline conditions, chelated iron, zinc, manganese and copper is a better way to provide micronutrients to a crop.
7. Chelates are compatible with a wide variety of pesticides and liquid fertilisers, as chelates do not react with their components. Most chelates can be mixed with dry mixes and liquid fertilisers.
8. Chelates are not readily leached from the soil as they adsorb on to the surface of soil particles.

Dr. S A Patil
Former Vice Chancellor UAS Dharwad,
Former Director of the IARI and
Former chairman Karnataka Krishi Mission
IMMA Souveneir, 2020

3.3 NANO-FERTILISERS

Nanotechnology is developing as the sixth revolutionary technology in the current era. It is considered as an emerging field of science widely subjugated in many scientific fields and supposed to play the main role in the field of agriculture and food science in next era. Nanotechnology applications have the potential to change agricultural production by allowing better management and conservation of inputs of plant and animal production. Nanotechnology provides a great scope of novel applications in the plant nutrition fields to cater to the future demands of the rising population (Waleed, 2018).

Nanotechnology is the science and technology of tiny things, the materials that are less than 100 nm in size. The term "Nanotechnology" was first defined in 1974 by Norio Taniguchi of the Tokyo Science University. By and large nanotechnology deals with structures in the size range between 1 to 100 nm and involves developing materials or devices within that size. Nanoparticles overlap in size with colloids, which ranges from 1 nm to 1 mm in diameter (Banfield and Zhang, 2001). However, the physical properties of nanoparticles are different from the properties of the bulk material (Buffle, 2006).

Agriculture is considered the backbone of most developing countries, with more than 60% of the population dependent on it for their livelihood. Modern agriculture involves the use of, among others, a substantial amount of inorganic fertilisers - a greater portion of which is removed from the realm of soil once the crop is harvested. Globally, crop yields have increased by at least 30 to 50% as a result of fertilization (Stewart et. al., 2005). Agricultural development has provided the much needed evidence that fertiliser application is the most efficient measure for substantially increasing crop production and ensuring food security (Bockman et. al., 1990) and that sustained yield growth is difficult without fertiliser supply (Larson and Frisvold, 1996).

World agricultural cropping systems intensively use large amount of fertilisers, pesticides, herbicides to achieve more production per unit area but also use more doses than optimum of these chemicals and fertilisers leading to several problems like environment pollution (soil, water, air pollution), low input use efficiency, decrease quality of food material, develop resistance in different weeds, diseases, insects, less income from the production, soil degradation, deficiency of micro nutrient in soil, toxicity to different beneficial living organism present above and below the soil surface etc. Efforts have been made in the past and are being tried at present to overcome this problem of fertiliser use. Many approaches have been made to increase the fertiliser use efficiency and reduce the dose. Among them the notables are: application of adequate amount of fertiliser (s); deep placement of fertiliser (s); use of granular urea; improving crop response knowledge (Brady and Weil, 2005) and use of slow release nano fertiliser (Ahmed *et. al.*, 2012).

Despite these problems there is also challenge to feed the growing population of the world. Plant nutrition is crucial for agriculture production and crop quality, and about 40% to 60% of the total world food production depends on the application of fertilisers. There is also the need to produce nutritive agricultural produce rich in protein and other essential nutrient required for human and animal consumption which is why emphasis is laid on the production of high quality food with the required level of nutrients and proteins.

NANOTECHNOLOGY APPLICATIONS IN AGRICULTURE

Nanotechnology plays an important role in agriculture by providing different nano devices and nano material which have a unique role such as nano biosensors to detect moisture content and nutrient status in the soil and also applicable for site specific water and nutrient management, Nano-fertilisers for efficient nutrient management, Nano-herbicides for selective weed control in crop field (Chinnamuttu and Kokiladevi, 2007), Nano-nutrient particles to increase seed vigor, Nano-pesticides for efficient pest management. alginate/ chitosan nano-particles can be use as herbicide carrier material specially for herbicide (Silva *et. al.*, 2011). Nanotechnology is widely used in Crop improvement, to increase efficiency of fertilisers and pesticides, soil management, plant disease detection, water management, analysis of gene expression and regulation and post-harvest technology (Waleed, 2018). There are naturally occurring nanoparticles that have been previously proposed for agricultural use, such as zeolite minerals. However, engineered nanomaterials can now be synthesized with a range of desired chemical and physical properties to meet various applications.

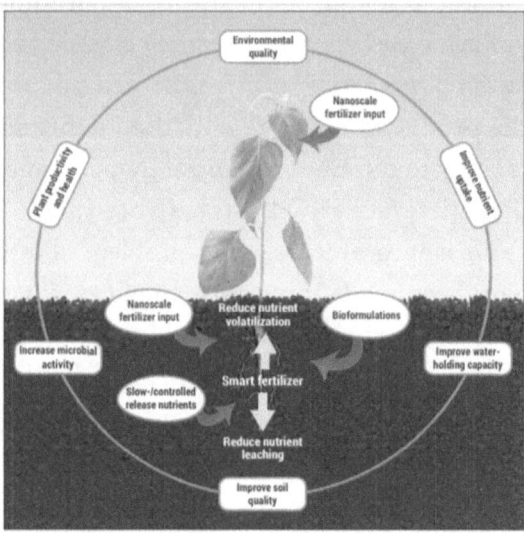

Fig. 3.3: Potential smart fertiliser effects in the soil-plant system. Adapted from Calabi-Floody et. al. 2017.

Nano-fertilisers

Nano fertilisers are synthesized or modified form of traditional fertilisers, fertilisers bulk materials or extracted from different vegetative or reproductive parts of the plant by different chemical, physical, mechanical or biological methods with the help of nanotechnology used to improve soil fertility, productivity and quality of agricultural produces. Nanoparticles can made from fully bulk materials (Brunnert, et. al., 2006).

Three classes of nanofertilisers have been proposed:

1. nanoscale fertiliser (nanoparticles which contain nutrients),
2. nanoscale additives (traditional fertilisers with nanoscale additives), and
3. nanoscale coating (traditional fertilisers coated or loaded with nanoparticles)

The increased efficiency of nanofertilisers is due to its reduced size and high surface area. This also enhances the reactivity and solubility of the fertilisers. The reduced size also facilitates more penetration of nano particles in to the plant from applied surface such as soil or leaves. Nanofertilisers have been advocated owing to higher NUE as plants cell walls have small pore sizes (up to 20 nm) which result in higher nutrient uptake (Fleischer and. Ehwald, 2014). Plant roots which act as the gateways for nutrients, have been reported to be significantly porous to nanomaterials compared to conventional manuring materials. The uptake of nanofertilisers can be improved by utilizing root exudates and molecular transporters through the ionic channels and creation of new micro-pores (Rico et. al., 2011). Nano-pores and stomatal openings in leaves have also been reported to felicitate nanomaterials uptake and their penetration deep inside leaves. Nanofertilisers have also been supported to have higher NUE owing to higher transport and delivery of nutrients through plasmodesmata which are nanosized (50–60 nm) channels for transportation of ions between cells (Zambryski, 2004).

NANOSCALE FERTILISERS AND THEIR FORMULATIONS

Different fertilisers inputs have been reported to be resized into smaller fractions through mechanical means or by employing specific chemical methods, which may increase nutrients uptake and reduce losses as well as nutrient toxicity. Nano-sized particles have been prepared from urea, ammonia, peat and other synthetic fertilisers as well as plant wastes. Mechanical cum biochemical approach is being employed to prepare such nanofertilisers where materials

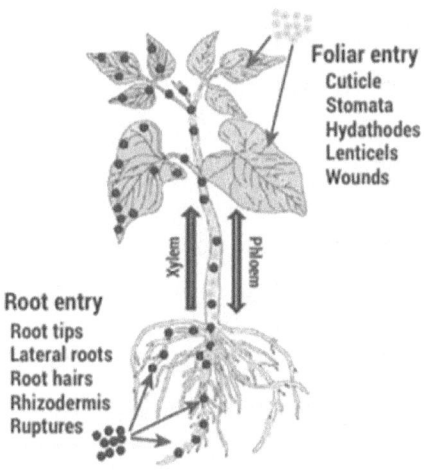

Fig. 3.4: Potential entry points of nanoparticles into plants. Wang et. al. 2016.

are grinded to nanosized particles through mechanical means and then biochemical techniques are put in action to prepare effective nanoscale formulations.

Nanomaterial coatings (such as a nanomembrane) may slow the release of nutrients or a porous nanofertiliser may include a network of channels that retard nutrient solubility. Fertilisers encapsulated in nano-particles increase the availability and uptake of the nutrient to the crop plants (Tarafdar, *et. al.*, 2013). Zeolite based nano-fertilisers are capable to release nutrient slowly to the crop plant which increase availability of nutrient to the crop though out the growth period which prevent loss of nutrient from denitrification, volatilization, leaching and fixation in the soil.

Another promising application of nanotechnology is the encapsulation of beneficial microorganisms that can improve plant root health. These could include various bacteria or fungi that enhance the availability of nitrogen, phosphorus, and potassium in the root zone. The development of nanobiosensors to react with specific root exudates is also being explored.

NANOFERTILISERS ADVANTAGES OVER CONVENTIONAL MINERAL FERTILISERS

Mineral nutrients if applied to crops in the form of nanofertilisers hold potential to offer numerous benefits for making the crop production more sustainable and eco-friendly (Subramanian, *et. al.*, 2015). Some of salient advantages are;

1. Nanofertilisers feed the crop plants gradually in a controlled manner in contradiction to rapid and spontaneous release of nutrients from chemical fertilisers.
2. Nanofertilisers are more efficacious in terms of nutrients absorption and utilization owing to considerably lesser losses in the form of leaching and volatilization.

3. Nanoparticles record significantly higher uptake owing to free passage from nano sized pores and by molecular transporters as well as root exudates. Nanoparticles also utilize various ion channels which lead to higher nutrient uptake by crop plants. Within the plant, nanoparticles may pass through plasmodesmata that results in effective delivery on nutrient to sink sites.
4. Due to considerably small losses of nanofertilisers, these can be applied in smaller amounts in comparison to synthetic fertilisers which are being applied in greater quantities keeping in view their major chunk that gets lost owing to leaching and emission.
5. Nanofertilisers offer the biggest benefit in terms of small losses which lead to lower risk of environmental pollution.
6. Comparatively higher solubility and diffusion impart superiority to nanofertilisers over conventional synthetic fertilisers.
7. Smart nanofertilisers such as polymer coated fertilisers avoid premature contact with soil and water owing to thin coating encapsulation of nanoparticles such as leading to negligible loss of nutrients. On the other hand, these become available as soon as plants are in position to internalize the released nutrients.

FIELD EVIDENCES OF NANOFERTILISERS USE FOR SUSTAINABLE CROPS PRODUCTION

The research findings of a field investigation proved in line with the postulated hypothesis where nano nitrogen fertilisers proved instrumental in boosting the productivity of rice. It was inferred that nano nitrogen fertiliser hold potential to be used in place of mineral urea and it can also reduce environmental pollution caused by leaching, de-nitrification and volatilization of chemical fertilisers (Milani et. al., 2012). Similarly, exogenously applied nutrients as nanomaterials increased the vegetative growth of cereals including barley (Rubio-Covarrubias et. al., 2009) while in contrast, nanofertilisers applied in conjunction with reduced doses of mineral fertilisers were found to be instrumental in boosting yield attributes and grain yield of cereals (Benzon et. al., 2015). Nanofertiliser of zinc applied as ZnO was found to be instrumental in boosting peanut yield due to robust plant growth, increased chlorophyll content of leaves and significantly better root growth (Prasad *et. al.*, 2012). The growth and yield boosting impact of different nanomaterials is depicted in Table 3.3

Table 3.3 Impact of nanofertilisers on productivity of different crops under varying pedo-climatic conditions (Iqbal, 2019)

Nanofertilisers	Crops	Yield increment (%)
Nanofertiliser + urea	Rice	10.2
Nanofertiliser + urea	Rice	8.5
Nanofertiliser + urea	Wheat	6.5
Nanofertiliser + urea	Wheat	7.3
Nano-encapsulated phosphorous	Maize	10.9
Nano-encapsulated phosphorous	Soybean	16.7
Nano-encapsulated phosphorous	Wheat	28.8
Nano-encapsulated phosphorous	Vegetables	12.0–19.7
Nano chitosan-NPK fertilisers	Wheat	14.6
Nanochitosan	Tomato	20.0
Nanochitosan	Cucumber	9.3
Nanochitosan	Capsicum	11.5
Nanochitosan	Beet root	8.4
Nanochitosan	Pea	20.0
Nanopowder of cotton seed and ammonium fertiliser	Sweet Potato	16
Aqueous solution on nanoiron	Cereals	8-17
Nanoparticles of ZnO	Cucumber	6.3
Nanoparticles of ZnO	Peanut	4.8
Nanoparticles of ZnO	Cabbage	9.1
Nanoparticles of ZnO	Cauliflower	8.3
Nanoparticles of ZnO	Chick pea	14.9
Rare earth oxides nanoparticles	Vegetables	7-45
Nanosilver + allicin	Cereals	4-8.5
Iron oxide nanoparticles + calcium carbonate nanoparticles + peat	Cereals	14.8-23.1
Sulfur nanoparticles + silicon dioxide nanoparticles + synthetic fertiliser	Cereals	3.4-45

In agreement to these findings, it was also reported that nanofertilisers of zinc improved the seed production of vegetables (Singh *et. al.*, 2013). Similarly, nano carbon incorporated fertilisers effectively reduced the days to germination and promoted root development of rice seedling. It was inferred that nano-composites have the potential to promote vital processes such as germination, radicle and plumule growth and development (Liu *et. al.*, 2008). Another aspect of nanofertilisers was explored regarding crop cycle as nanoparticles which were loaded with NPK, reduced the crop cycle of wheat up to 40 days, while grain yield was also increased in comparison to mineral fertilisers applied at recommended rates (Abdel-Aziz *et. al.*, 2016). Slow release fertiliser coated with nanoparticles boosted the productivity of wheat-maize cropping system (Xiao *et. al.*, 2008). In addition to soil applied nanofertilisers, foliar application of chitosan was reported to be instrumental in boosting tomato yield by 20%, while it remained non-significant as far as carrot yield was concerned (Walker *et. al.*, 2004). However, growth promoting effect of foliar applied chitosan was also recorded for horticultural crops such as cucumber, beet-root etc. The significantly higher selenium uptake by many crops including green tea was observed when it was applied as nanosized particles (Yoo *et. al.*, 2009). There are various other impacts that can be imparted by nanomaterials in different crops and some of these have been described in Table 3.4

Table 3.4 Characteristics imparted by nanomaterials in different crops (Iqbal, 2019)

Nanofertilisers	Crops	Imparted Characteristics
Nanoparticles of ZnO	Chickpea	Increased germination, better root development, higher indoleacetic acid synthesis.
Nano silicon dioxide	Maize	Drought resistance, increment in lateral root roots number along with and shoot length.
Nano silicon dioxide	Maize	Increased leaf chlorophyll.
Nano silicon dioxide	Tomato	Taller plants and increased tuber diameter.

Colloidal silica + NPK fertilisers	Tomato	Increased resistance to pathogens.
Nano-TiO$_2$	Spinach	Improved vigor indices and 28% increased chlorophyll.
Polyethylene + indium oxide	Vegetables	Increased sunlight absorption
Polypropylene + indium–tin oxide	Vegetables	Increased sunlight utilization
Gold nanoparticles + sulphur	Grapes	Antioxidants and other human health benefits.
Kaolin + SiO$_2$	Vegetables	Improved water retention
Bentonite + N-fixing bacteria inoculation	Legumes	Improved soil fertility and resistance to insect pest
Nanocarbon + rare earth metals + N fertilisers	Cereals	Improved nitrogen use efficiency
Stevia extract + nanoparticles of Se + organo-Ca + rare-earth elements + chitosan	Vegetables	Enhanced root networking and root diameter
Nano-iron slag powder	Maize	Reduced incidence of insect pest
Nano-iron + organic manures	Cotton	Controlled release of nutrients acts as an effective insecticide and improves soil fertility status.

LIMITATIONS OF NANO FERTILISERS

Despite offering numerous benefits pertaining to sustainable crop production, nanofertilisers have some limitations regarding research gaps, absence of rigorous monitoring and lack of legislation which are currently hampering the rapid development and adoption of nanoparticles as a source of plant nutrients (Remedios *et. al.*, 2012). A few of the limitations and drawbacks associated to nanofertilisers use for sustainable crop production are enlisted below.

1. Nano fertilisers related legislation and associated risk management continue to remain the prime limitation in advocating and promoting nanofertilisers for sustainable crop production.
2. Another limiting factor is the production and availability of nano fertilisers in required quantities and this is the foremost limitation in wider scale adoption of nanofertilisers as a source of plant nutrients.
3. The higher cost of nanofertilisers constitutes another hurdle in the way of promulgating them for crop production under varying pedo-climatic conditions across the globe.
4. Another major limitation pertaining to nanofertilisers is the lack of recognized formulation and standardization which may lead to contrasting effects of the same nanomaterials under various pedoclimatic conditions.
5. There are many products being claimed to be nano but in fact are submicron and micron in size. This dilemma is feared to remain persistent until and unless uniform size of nanoparticles (1–100 nm) gets implemented.

Nanofertilisers applied alone and in conjunction with organic materials have the potential to reduce environmental pollution owing to significant less losses and higher absorption rate. In addition, nanomaterials were recorded to improve germination rate, plant height, root development and number of roots, leaf chlorophyll and fruits antioxidant contents. Moreover, controlled and slow released fertilisers having coating of nanoparticles, boost nutrient use efficiency and absorption of photosynthetically active radiation along with considerably lower wastage of nutrients.

The future of nanofertilisers for sustainable crop production and time period needed for their general adaptation as a source of plant nutrients depend on varied factors such as effective legislation, production of novel nanofertilisers products as per requirement and associated risk management. There is a dire need for standardization of nanomaterials formulations and subsequently conducting rigorous field and greenhouse studies for performance evaluation. For sustainable crop production, smart nanofertilisers having the potential to release nutrients as per plants requirement in temporal and spatial dimensions must be formulated. Lastly, researchers and regulators need to shoulder the responsibility by providing further insights in order to take full advantage of the nanofertilisers for sustainable crop production under changing climate with the risk of causing environmental pollution.

Nanotechnology in The World Of Agriculture

Nanotechnology is the manipulation or self-assembly of individual atoms, molecules, or molecular clusters into structures to create materials and devices with new or vastly different properties. It is defined as "the understanding and control of matter at dimensions of roughly 1-100 nm, where unique properties make novel applications possible". By manipulating molecules on the scale of billionths-of-a-meter, scientists have created materials that exhibit "almost magical feats of conductivity, reactivity, and optical sensitivity, among others". Nanotechnology has wide applications in medicine (Zhou et. al., 2004), environment (Shi et. al., 2007), energy (Das et. al., 2007), information and communication (Hillie, 2007), heavy industry (Lo et. al., 2010) and consumer goods (Schneider, 2010).

Nano and the Agri-food industry

Despite the fact that nanotechnology is extensively exploited in health and medical sciences, the use of nanotechnology in agriculture and food sector is just emerging and expected to grow alarmingly. Globally, agrifood sector has been identified as a potential industry to make significant investments. Currently, over 300 nanofood products are available in the international markets. Retailers already sell over 300 products incorporated with nanotechnology (Science Policy Council, 2007). In 2016, about 14,000 documents with nanotechnology in food or agriculture were listed meaning high activities of this field. Also about 2707 patents matched this criterion are found in world patent database (patentscope.wipo.int). The Food and Agriculture sectors also require

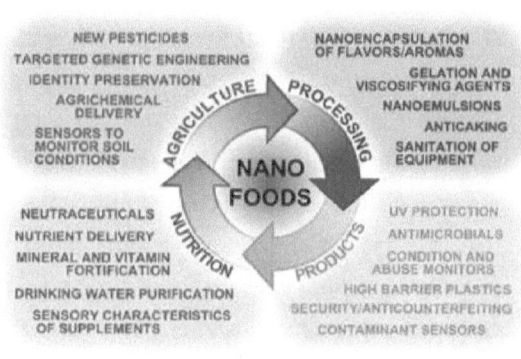

high amount of nanomaterials especially in packaging. The NAFTA region shares the biggest slice from the market size, but Europe and Asia especially China, Japan, and India have also come up very dynamically.

Need for nanoagri-inputs

Farmers across the globe are left with the daunting task of feeding more mouths every year from agricultural fields which are dwindling correspondingly. The production level of food grains has become a subject of concern as it has been showing a downward trend over the last decade. Nutrient use efficiencies of conventional fertilisers hardly exceed 30-35 %, 18-20 % and 35-40 % for N, P and K respectively. Also, the use of agrochemicals is associated with some risks for human and environmental health (e.g., contamination of water resources, residues on food products).

Nanotechnology has the potential to revolutionize the agricultural and food industry with new tools for the molecular treatment of diseases, rapid disease detection, enhancing the ability of plants to absorb nutrients etc. Nanofertilisers are nutrient carriers that are being developed using substrates with nanodimensions of 1 – 100 nm. Nano particles have extensive surface area and capable of holding abundance of nutrients and release them slowly and steadily such that it facilitates uptake of nutrients matching the crop requirement without any associated ill effects of customized fertiliser inputs thereby improving the fertiliser use efficiency of crops. Smart sensors and smart delivery systems will help the agricultural industry combat viruses and other crop pathogens. In the near future nanostructured catalysts will be available which will increase the efficiency of pesticides and herbicides, allowing their usage at lower doses. It is noteworthy to mention that nanomaterials could also be exploited to improve structure and function of pesticides by increasing solubility, reducing hydrolysis and photodecomposition, and/ or by providing a more specific and controlled-release toward target organisms (Mishra and Singh, 2015; Grillo et. al., 2016; Nuruzzaman et. al., 2016). Nanotechnology will also protect the environment indirectly *through the use of alternative (renewable) energy supplies, and filters or catalysts to reduce pollution and cleanup the existing pollutants.*

Novel nanomaterials based on the use of inorganic, polymeric, and lipid nanoparticles, synthesized by exploiting different techniques (e.g., emulsification, ionic gelation, polymerization, oxydoreduction, etc.) are being developed to increase productivity. They can find application in the development of intelligent nanosystems for the immobilization of nutrients and their release in soil. Such systems have an advantage to minimize leaching, while improving the uptake of nutrients by plants, and to mitigate eutrophication by reducing the leaching of nitrogen to groundwater (Liu and Lal, 2015).

Hydrogels, nanoclays and other minerals have been reported to enhance the water-holding capacity of soil (Sekhon, 2014), hence acting as a slow release source of water, reducing the hydric shortage periods during crop season. Organic e.g., such as polymer and carbon nanotubes and inorganic e.g., such as nano metals and metal oxides nanomaterials have also been used to absorb environmental contaminants (Khin et. al., 2012), increasing soil remediation capacity and reducing time and cost of the treatments.

Many reports foresee that nanotechnology will allow the development of high-tech agricultural fields, equipped with a range of intelligent nanotools that allow the precised management and control of inputs, including pesticides, fertilisers, and water. The development of such devices would certainly lead to a revolution in agricultural practices, and could possibly contribute to an important reduction in the impact of modern agriculture on the environment and an improvement in both the quality and quantity of yields (Scott and Chen, 2002; Sekhon, 2014; Liu and Lal, 2015).

Nanotechnology Applications in Agriculture

Mentioned below are the brief applications of Nanotechnology in the field of Agriculture.

1. *Nano-formulations of agrochemicals for applying pesticides and fertilisers for crop improvement and plant nutrition.*
2. *The application of nanosensors/ nanobiosensors in crop protection for the identification of diseases and residues of agrochemicals.*
3. *Nano-devices for the genetic manipulation of plants.*
4. *Animal health, animal breeding, poultry production.*
5. *Post-harvest management.*
6. *Increase the efficiency and food quality due to accelerated absorption.*
7. *Prevention in the rate of fertilisers loss either by leaching and or fast uptake of nutrients by plants and allows controlled release of nutrients in the growth period.*
8. *Reduction in soil and water pollution and consequently food products through reduced leaching of fertilisers.*
9. *Increase nutrients use efficiency, reduce soil toxicity, and minimize the potential negative effects associated with different over dosages (Solanki et. al. 2015; Belal and El-Ramady 2016; Chhipa and Joshi 2016; Chhipa2017; Khan and Rizvi 2017; Subramanian and Thirunavukkarasu 2017).*

When considering all nanoproducts that will possibly emerge in the food and agriculture sectors, there are widely accepted consensus that there is a need to gather more reliable data currently to allow a clear safety assessment (FAO/ WHO, 2013; JRC-IPTS, 2014). When considering only nanoagrochemicals, the paradigm behind a classical risk assessment approach (i.e., hazard × exposure) is suitable, but applying approaches used within the current regulatory framework directly would result in a number of pitfalls (Kookana et. al., 2014). Exposure assessment relies on investigations into the environmental fate of a compound. There have been a limited number of studies investigating nanoagrochemicals (Kah et. al., 2013; Kah and Hofmann, 2014). Overall, the current level of knowledge needs to be enhanced for a reliable assessment of the risks associated with the use of nanoagrochemicals.

Nanotechnology will soon become an integral part of any nation's future. Research in nanotechnology can do a lot to benefit society through applications in agriculture and food sectors (Sugunan and Dutta, 2004) Introduction of any new technology always has an ethical responsibility associated with it to be apprehensive to the unforeseen risks that may come along with the tremendous positive potential. Public awareness about the advantages and challenges of nanotechnology will lead to better acceptance of this emerging technology. Nanotechnology applications in agriculture and food systems is still at the nascent stage and a lot more applications can be expected in the years to come.

*- **Dr. Shama Zaidi**
Sr. Manager (R&D),
Aries Agro Ltd
IMMA Souveneir,2018*

Integrated Nutrient Management

4.1 IPNM- INTRODUCTION

The use of organic manures as source of nutrients dates back to the beginning of settled agriculture but after the introduction of widespread use of mineral fertilizers, organic manures were thought of as a secondary source of nutrients. However, with increasing awareness about soil health and sustainability in agriculture, organic manures and many diverse organic materials, have gained importance as components of integrated plant nutrient management (IPNM) strategies. The basic concept underlying IPNM is the maintenance and possible improvement of fertility and health of the soil for sustained crop productivity on long-term basis and use fertilizer nutrients as supplement to nutrients supplied by different organic sources available at the farm to meet the nutrient requirement of the crops to achieve a defined yield goal.

IPNM concentrates on a holistic approach to optimizing plant nutrient supply. It includes the following considerations:

- Assessing residual soil nutrient supplies, as well as acidity and salinity;
- Determining soil productivity potential for various crops through assessment of soil physical properties with specific attention to available water holding capacity and rooting depth;

- Calculating crop nutrient requirements for the specific site and yield objective;
- Quantifying nutrient value of on-farm resources such as manures and crop residues;
- Calculating supplemental nutrient needs (total nutrient needs minus on-farm available nutrients) that must be met with "off-farm" nutrient sources;
- Developing a program to optimize nutrient utilization through selection of appropriate nutrient sources, application timings and placement.

The overall objective of IPNM is to adequately nourish the crop as efficiently as possible, while minimizing potentially adverse impacts on the environment. The trend globally is to find a balance of nutrient sources that takes advantage of the recycling of nutrients in manures and crop residues, supplementing them with commercial fertilizers. If this requires processing and transportation of the organic sources, those expenses must be considered. But it is also important to consider the costs and environmental consequences of not finding a way to utilize the organic materials. Recycling of available organic nutrient sources should be included in the nutrient planning whenever it is practical.

Fig 4.1: Performance applying farmyard manure (FYM) and fertilisers as integrated plant nutrient management vis-à-vis fertiliser alone in rice

During last 2 to 3 decades, IPNM has emerged as an effective agricultural paradigm to ensure food security through increased production of rice and wheat, particularly in countries with developing economies. A large number of investigations in these regions have shown a significant positive impact on yield of rice and wheat through integrated management of different organic materials and fertilisers.(*FARMING OUTLOOK September Issue 2018*)

POSITIVE EFFECT OF IPNM ON DIFFERENT CROPS:

In recent years, besides rice and wheat, positive effect of IPNM in terms of yield has been recorded in cotton, maize, potato, barley, lentil, sunflower, finger millet, mustard, kinnow and several vegetable crops. In semi-arid tropics too, IPNM practices have been found superior to mineral fertilisers alone in realizing either at par or higher yield levels of chickpea, wheat, soybean, pearl millet, maize and groundnut and substituting the use of 50% of mineral fertilisers through effective recycling of on-farm wastes. Wheat is grown in winter season so that mineralization of organic sources of nutrients is slow as compared to in the summer when rice and maize are grown in South Asia.

Therefore, IPNM in wheat is less popular than in rice or maize. Although in some studies positive effect of IPNM on wheat yield has been reported, substantial residual effect of organic materials applied to preceding summer season crops of rice, maize or soybean has been observed in wheat by several researchers. *FARMING OUTLOOK September Issue 2018)*

LONG-TERM EXPERIMENTS CONDUCTED IN INDIA

In several long-term experiments on rice-wheat cropping system conducted for 12 to 15 years at different locations in India (Yadav *et. al.*, 2000), farmyard manure (FYM), GM and CR were applied in combination with NPK fertilisers to rice whereas to wheat only fertiliser NPK was applied. When pooled data across the locations were used to fit linear regressions, highly significant annual increase in yield of both rice and wheat with IPNM treatments was observed, indicating thereby the advantage of combined use of manures plus fertilisers over fertilisers alone in sustaining crop yields. In another 12 long-term fertility experiments on rice-wheat cropping system in the Indo-Gangetic plains of South Asia (Ladha *et. al.*, 2003), average yield change in rice was significantly higher with the application of FYM and GM along with 50%

NPK as compared with the only 100% NPK treatment.

MITIGATION OF GLOBAL WARMING

Nitrous oxide is an important greenhouse gas contributing to global climate change. It is produced through soil microbial processes of nitrification and denitrification, and contributes to roughly 6% of the overall radiative forcing in the atmosphere. Addition of easily decomposable organic C in the form of organic amendments may increase microbial activities and induce anaerobic conditions, which may lead to increased losses of nitrous oxide via denitrification. But in several investigations under both upland crops and lowland rice, integrated application of inorganic fertilisers and organic sources of nutrients resulted in reduced or no significant effect on nitrous oxide. A few studies reported increased emission but the trends were obviously linked not only to the quantity and quality of organic amendments but also to soil biotic and abiotic factors.

Successful implementation of IPNM requires a team effort among all stakeholders:

- *Policy makers* are needed to provide funding for research and extension activities and support for training, research, data management, and advisory activities.
- *Research institutions* provide the local science to adapt practices, develop tools for implementing and monitoring results, analyze and interpret data collected, and provide educational programs to improve the decision process.
- *Extension and agribusiness* dealers are the front-line contact with the farmers and help provide guidance and answer questions about adapting technology and practices to the local conditions and culture.

- *Fertilizer manufacturers* play an important role in supplying the right products for each area in sufficient quantities and at the right time. They support local research and training for local input suppliers, advisers, and farmers.
- *Farmers* may be the most important members of the team. They make the final decisions and take the final steps to implement IPNM, assess the final result, and reap the rewards of successful IPNM implementation.

Fertilisers have played a vital role in increasing crop productivity. However, environmental issues have been linked with indiscriminate use of fertilisers. Combined use of organic nutrient sources and fertilisers following the concepts of IPNM guarantees optimal use of fertilisers to achieve high crop production levels but with minimal adverse effects on the environment.

4.2 4R NUTRIENT STEWARDSHIP FOR FERTILIZER MANAGEMENT

There are four management objectives associated with any practical farm level operation, including management of fertilizers. These are productivity, profitability, cropping system sustainability, and a favourable biophysical and social environment. Best management practices for fertilizer support the realization of these objectives in terms of cropping and the environmental health. 4R nutrient stewardship requires the implementation of best management practices (BMPs) that optimize the efficiency of fertilizer use. The goal of fertilizer BMPs is to match nutrient supply with crop requirements and to minimize nutrient losses from fields.

A strong set of scientific principles guiding the development and implementation of fertilizer best management practices (FBMPs) has evolved from a long history of agronomic and soil fertility research. A *Global Framework for Fertilizer Management* has been developed and is being adopted to guide nutrient stewardship. While this system has not yet been adopted in all parts of the world, it provides a good outline of the interactions of the scientific, economic, and social aspects of nutrient management. Described as *"4R Nutrient Stewardship"* (Table 4.1), it provides a framework to achieve cropping system goals of increased production, increased farmer profitability, enhanced environmental protection, and improved sustainability. To achieve those goals, the 4R concept incorporates the:

Right fertilizer source at the
Right rate, at the
Right time and in the
Right place

Table 4.1 Components of the 4R Nutrient Stewardship system.

Component	Goal
Right source	Provide plant-available forms, and a balanced supply of all essential nutrients. Take advantage of various formulations that offer improved efficiency and reduce environmental consequences.
Right rate	Ensure an adequate supply of all essential nutrients to meet plant demand.
Right time	Manage nutrient applications to match the interactions of crop uptake, soil supply, environmental risks, and field operation logistics.
Right place	Consider root-soil dynamics and nutrient movement, and manage spatial variability within the field to meet site-specific crop needs and minimize potential losses from the field.

Selection of FBMPs varies by location, and those chosen for a given farm are dependent on local soil and climatic conditions, crop management conditions and other site-specific factors.

For each of the 4R components, a series of performance indicators related to the economic, environmental and social goals have been identified to serve as measures of performance.

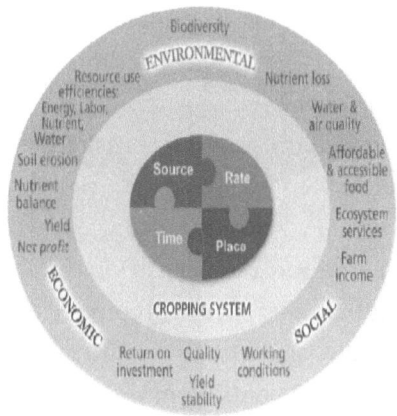

Fig. 4.2: *Economic, social and environmental goals related to nutrient management (IPNI, 2012).*

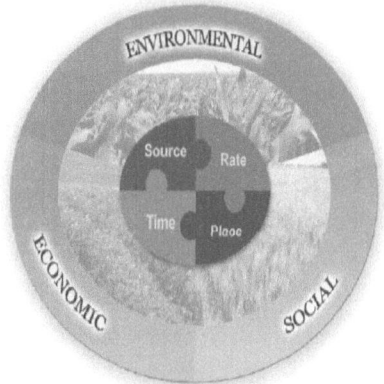

Fig. 4.3 *Performance indicators (IPNI, 2012).*

4R IMPLEMENTATION FOR SUSTAINABLE CROPPING SYSTEMS

Production demands, input requirements and environmental impacts taken together mean the risks for making the wrong nutrient use decisions is greater now than ever. When fertilizer BMPs result in increased production and input use efficiency, they also reduce losses to the environment. When making practice selection, the interconnectivity between practices addressing source, rate, time and place should be considered. Performance measures and indicators will often include crop yields and sufficient information to calculate economic returns. In addition, they will need to reflect environmental and social performance. Those chosen may vary depending on stakeholder priorities, but will often include either nutrient balances or nutrient use efficiencies. Crop advisors are key in the efforts to increase adoption of 4R Nutrient Stewardship with growers. Fertilizer BMPs should be selected based on these principles, and should then be used in combination with other conservation practices.

Right source

The right source for a nutrient management system must provide a balanced supply of all essential nutrient elements in plant-available forms. The right source must also consider any nutrient interactions or compatibility issues, potential sensitivity of crops to the source, and any non-nutrient elements included with the source material. The right source may vary with the crop, the soil properties of the field, and options for method of application. These include fertilizer products of various kinds that slow chemical conversions, encapsulate fertilizer materials in some kind of protective coating or in other ways modify the rate or release of the nutrients from the fertilizer materials.

Right rate

Soil testing and plant analysis are important tools to help with the decisions regarding the supplying power of the soil in relation to the nutrient requirement of the crop. Plants require different rates of different nutrients at different stages of the growing season. Rate should be adjusted to help balance nutrient supply with crop removal at all times to avoid deficiency stress and economic loss. Excessive rates may lead to inefficiency in nutrient use and economic losses and environmental problems. In some cases excess nutrients may also result in toxicity to the crop. With precision farming technologies, such as GIS-referenced soil testing, variable-rate fertilizer application, and harvest with yield monitors, on-farm rate studies can be easily done and can guide site-specific, variable-rate fertilizer application to provide the most efficient and most profitable fertilizer nutrient management system, applying different rates of fertilizer on different areas within an individual field.

Right time

Crop nutrient requirements change throughout the growing season as the crop grows from vegetative stages, through reproductive stages, and on to maturity. The slow- and controlled-release and enhanced efficiency fertilizer products and additives help to provide a broader choice of options for timing the nutrient availability to crop requirements, and thus options for the time and method of application. One of many examples of timing fertilizer applications based on stage of crop growth and nutrient needs is *split-application of N*. For example, a small amount of N may be surface applied as UAN solution in the fall to stimulate soil micro-organisms and help support decomposition of previous crop residues. A second, pre-plant, application using banded anhydrous ammonia or UAN solution may then provide the majority of the N requirement, followed by a supplemental side-dress or top-dress application to fine-tune the total N program based on in-season monitoring, or predetermined total N rate plans

Right place

Having the nutrients in the right place helps to ensure that plant roots can absorb enough of each nutrient at all times during the growing season.

For placement with respect to the seed row and growing plant roots, there exist several options:

- Surface broadcast or band application
- Starter fertilizer application with the traditional 5 cm x 5 cm placement
- Deeper banding (usually 10 to 15 cm below the surface, providing a concentrated nutrient source lower in the root zone).

A narrow strip (about 1/3 of the surface) is tilled and nutrients are concentrated in a band below the surface, maintaining a predominantly untilled surface residue environment to help reduce erosion and conserve soil moisture. The right place also depends upon the characteristics of the fertilizer material being applied. Fertilizers applied to the soil surface are subject to potential losses in surface runoff. Other materials, such as urea or UAN solution, may be surface applied, but volatilization losses can be substantial without sufficient rainfall within a few days to move the fertilizer into the soil.

With precision farming tools, variability in nutrient needs based upon soil tests and other yield potential factors can be met with variable-rate application to match fertilizer applied to varying crop needs on a site-specific basis within the field. The placement of fertilizer affects both the current crop and subsequent crops.

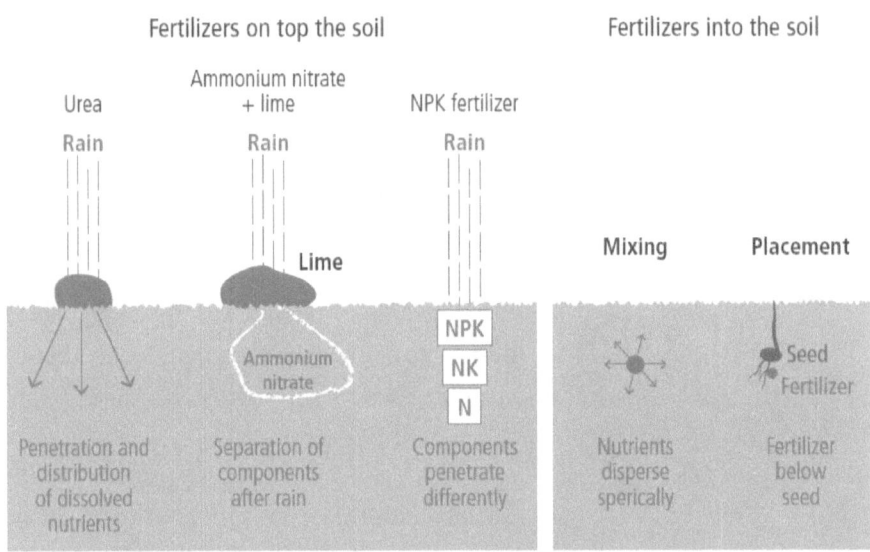

Fig. 4.4: Impact of different fertilizer placement practices for movement of nutrients into the soil (adapted from IFA, 1992).

Broadcast application over time results in a uniform distribution of nutrients, which gradually move down the soil profile deeper into the root zone. Band application in the same location over time accounts for specific variability of soil test levels, and in relationship to the growing roots with precision guidance and placement systems. Strip-till systems are especially useful in conjunction with Real Time Kinematic (RTK) guidance systems to ensure the fertilizer band is placed in an exact relationship with the seed row, even though the fertilizer may be applied several months in advance of planting. With RTK guidance systems, farmers can apply starter fertilizer in the fall, and then plant the seeds the following spring, with the seed row accurately placed in the desired relationship with the starter fertilizer band. Thus, RTK provides accuracy in placement of fertilizer wherever it is needed, as well as providing options for timing of application with respect to the cropping season. Band application, without controlled guidance, results in multiple bands, and over time approximates the effect of broadcast application.

Integrated Nutrition Management for Alleviation of Abiotic Stress In Crops

To meet food demands of a growing human population, the productivity of agricultural land and soil need to be improved. So as to enhance crop productivity, eco-friendly strategy for sustainable agricultural development require priority policies in India.

Prior to green revolution, the farming scenario was different. Farmers used to cultivate single crop with sustainable production. Farmers used to follow conventional and traditional methods for cultivation and land was kept fallow, organic manures were used, thus traditional farming methods helped to replenish the soil fertility.

In order to promote food security, Green revolution was introduced and has created revolutionary and significant growth in food production. Farming scenario was modified by acceptance of modern farming techniques, which includes high yielding varieties, high inputs of synthetic fertilizers and pesticides, greater exploitation of irrigation facilities (ground water and canals), mechanized agricultural equipments, which has helped to gain higher yields to overcome the increasing food demand.

With the green revolution of meeting the food demand and increasing the productivity, indiscriminate use of fertilizers and pesticides, irrigation facilities, low or no use of organic manure has impacted severely on soil health and environment and declined the productivity of crops in India. This devastating change also has increased abiotic stress, which has severely influenced on the ecological and agriculture systems.

PLANT STRESS

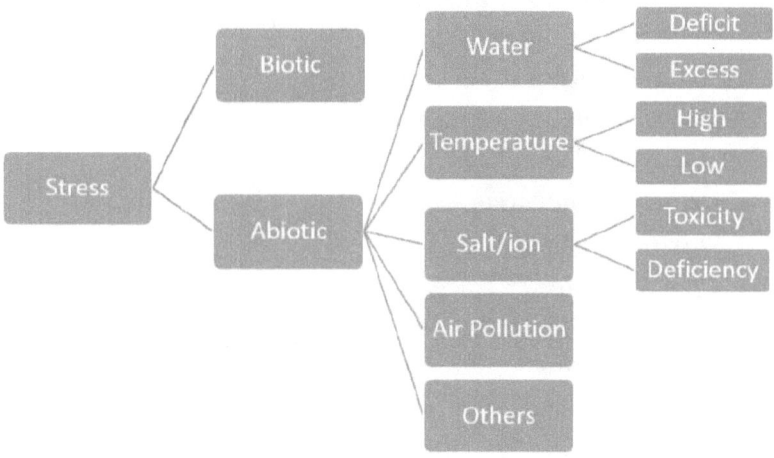

WHAT IS PLANT STRESS?

A sudden biology change in environmental conditions that might reduce or adversely affects plant's growth & development.

WHAT IS ABIOTIC STRESS?

Abiotic stress is defined as the negative impact of non-living factors on the living organisms in a specific environment.

Abiotic stress comprises of:

Drought & water logging, salinity, low & high temperatures, inadequate & excess mineral nutrients in soil, rainfall, floods, storms, wind, light & radiation stress & many other factors. Besides these factors such as global climatic changes has impacted crop production and need to be addressed on priority basis.

Impact of abiotic stress:

- Germplasm extinction
- Stagnation in crop yields

- *Multi nutrient deficiencies*
- *Shrinking agricultural lands*
- *Contamination of soil & water*
- *Declining soil fertility & productivity*
- *New spectrum of pest & diseases*
- *Physiological disorders in crops*
- *Poor seed germination*
- *Nutrient imbalance in soil*

In order to overcome this challenge, it is important to have an integrated nutrient management approach for meeting of climatic change effects on crops.

Integrated nutrient management (inm):

Integrated Nutrient Management is to maintain soil fertility & balance plant nutrient supply to an optimum level for sustaining the desired crop productivity (through optimization of benefits from all possible sources of plant nutrients in an integrated manner). Presently in India, the nutrient use efficiency status is quite low due to the deterioration of physical, chemical and biological properties of the soil and has drastically reduced crop productivity.

Integrated Nutrient Management has all-round potential to get better crop response, proper and efficient use of nutrient resources, so as to get better and higher yields and protecting from deterioration of soil health and other resources.

Integrated nutrient management aims:

- *To maintain soil productivity & its fertility through balanced plant nutrition in combination with organic & biological source of plant nutrients*
- *To improve the plant nutrient efficiency & reduce the losses to the environment*

- *To improve the physical condition of the soil by utilizing benefits of organic manures, green manures & bio-fertilizers*
- *To meet the socioeconomic aspects without harming natural resource base of agricultural production*

Integrated nutrient management motive is to improve soil health. So as to improve soil health, properties of soil such as physical, chemical and biological need to be addressed properly.

- *Physical properties of soil are important because it determines the anchorage of the plant, root penetration, aeration, drainage, moisture retention etc,.*
- *Chemical properties of soil includes cat-ion & an-ion exchange capacity, base exchange capacity, nutrient interaction, soil pH, Carbon:Nitrogen ratio (C:N), Electric conductivity (EC), buffering capacity etc., Chemical properties are governed by the applied nutrients and native nutrients available in the soil.*
- *Biological properties of soil comprises of large number of beneficial soil organisms range from bacteria (0.000001 mm) to giant tunnelling earthworms (1 m). Presence of these soil organisms and their various functions makes the soil a living entity.*

Components of integrated nutrient management:

- *Balanced Plant Nutrients (BPN)*
- *Organic fertilizers/ Green manures & compost*
- *Bio-fertilizers*
- *Plant Growth Promoters*

- *Balanced Plant Nutrition is the key 16 to 17 essential plant nutrient applied to in balanced form as per prophylactic growth stages of the crops. Absence or excess of any nutrient can affect the plant well being. If the deficient element is xsupplied, growth will be increased up to the point,* 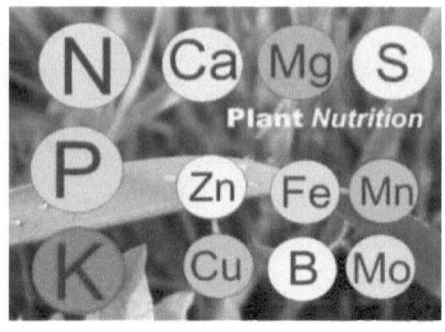 *where the supply of that element is no longer the limiting factor. Balanced plant nutrition ensure optimum yield from the crops by enabling them to reach their maximum genetic potential. It not only helps to increase yields, but also builds crop resistance against biotic & abiotic stress in crops.*

Merits of integrated nutrient management:

- *Avoids wastefulness of nutrients in form of fertilizers both in terms of nutrient & cost of cultivation*
- *Provides better yields by adding utilization of added nutrients efficiently*
- *Ensures optimum & continuous qualitative yield*
- *Negative interaction within the nutrients is reduced or nullified*
- *Imparts natural resistance against biotic & abiotic stress*
- *Builds resistance against adverse environmental condition*
- *Nullifies the ill effects of soil abnormalities & improves soil health*
- *Provides Cost: Benefit ratio*

To summarize:

The increasing food demands of growing population of India, there is a need for an environment friendly strategy for sustainable agricultural development, which requires significant attention, when address the issue of crop productivity.

One of the important agronomic solution to overcome abiotic stress approach of integrated nutrition management is essential. It can play a vital role in alleviating the stressful effects of abiotic stress in crops. Integrated nutrient management plays an important role in sustaining soil health and getting better yield from crops.

It is important and essential to impart proper training on successful "Soil Health Management and Integrated Nutrition Management" program to farming community, along with evolution of new tools and techniques to overcome the ill effects of abiotic stress by maintaining soil health for sustainable agriculture. This will help to meet food demands of a growing human population for the present and coming generations with smart agriculture approach in India.

Devendra D. Dange
Technical Director
Email: manshya@gmail.com
Manshya Marketing Pvt. Ltd., Pune
IMMA Souvenir. 2018

4.3 PRECISION FARMING AND SITE-SPECIFIC NUTRIENT MANAGEMENT

Developments in computer technology, geographic information systems (GIS), global positioning system (GPS), electronic sensors and controllers, and a wide variety of communication tools during the 1990s and into the 21st century have provided exciting new technologies that can be applied in agriculture. Under the collective term, precision farming, these technologies have opened many new opportunities for improving crop and soil management on a site-specific basis.

Precision farming is an approach where inputs are utilised in precise amounts to get increased average yields, compared to traditional cultivation techniques. Popular definitions of Precision Agriculture (PA), Satellite Farming or Site Specific Crop Management (SSCM) or Site Specific Nutrient Management (SSNM) describe the term as **'a technology-enabled approach to farming management that observes, measures, and analyzes the needs of individual fields and crops'**. While initially designed and developed for broad-acre, large-scale producers in the US, West Europe and South America, precision farming has many applications that fit equally well for the small farmer.

The development of precision agriculture is shaped by two trends: "Big Data and Advanced Analytics Capabilities, and Robotics — aerial imagery, sensors, sophisticated local weather forecasts". In simple words farming that collects and uses data of plots for managing and optimizing the production of crops is known as Predictive farming

In India, one major problem is the small field size. More than 58 per cent of operational holdings in the country have size less than one hectare (ha). Only in the states of Punjab, Rajasthan, Haryana and Gujarat do more than 20 per cent of agricultural lands have an operational holding size of more than four ha. Commercial as well as horticultural crops also show a wider scope for PA in the cooperative farms.

SITE SPECIFIC NUTRIENT MANAGEMENT

Site specific nutrient management (SSNM) is an approach of supplying plants with nutrients to optimally match their inherent spatial and temporal needs for supplemental nutrients. The SSNM provide an approach for need based 'feeding' of crops with nutrients.

SSNM fits anywhere in the world, and often may be easier to implement on small-scale farms where each field is more carefully monitored and managed. It is not limited to large fields and large equipment. The concept of SSNM attempts to match best management practices to the individual location, considering that location has unique soil and climate, and the unique management skills and experience of the grower.

The main features of SSNM are: (Tiwari, 2007)

Site specific application of nitrogen, phosphorus and potassium and secondary and micronutrients based on soil tests are followed.

- Optimal use of existing nutrients, such as from soil, residues and manures.
- SSNM further provides guidelines for selection of the most economic combinations of nutrients.
- Advocates wise and optimal use of existing indigenous nutrient sources such as crop residues and manures.

Site-specific management technology relies on the interaction of three broad and fundamental elements to be successful in its implementation. They are categorized in terms of information, technology and management.

Information

In field variability, spatially or temporally, soil related properties, crop characteristics, weed and insect pest population and harvest data are important databases that need to be developed to realize the potential of site-specific management technology. Out of these, crop yield monitoring is the most mature component and logical starting point. Several years of yield data may be required to make a good decision. Highly varying yield within field indicates that the current management practices may not be providing the best possible growing conditions everywhere in the field (Verma, *et.al.*, 2020).

Establishment of soil related characteristics within field, through regular soil sampling, is another database that is extremely important. Some of the characteristics such as soil texture vary very little over time, others such as moisture content and nitrate level, fluctuate rapidly. Decision therefore, has to be made on what property to sample, how to sample and how often to sample so that interpretation from database can be made with greater confidence (Verma, *et.al.*, 2020).

Technology

The recent development in microprocessor and other electronic technologies for monitoring yields and sensing soil related variables are new tools available to make site specific farming a success. When measuring soil characteristics such as the harvest data, moisture and nutrients availability in the soil, satellite based positioning system, namely, geographical positioning system (GPS) can be used to identify the locations where the data are taken. Some GPS users demand accuracy in identifying field location and differential global positioning systems (DGPS) is one of the improved GPS system that reduce position errors. With this information, the results of soil sampling test and yield data can be transformed in to field maps, achievable through personal computers (PC) and geographic information systems (GIS) software. The same map can be developed for other field characteristics such as weed and salinity mappings.

Remote sensing technique can also be utilized to detect soil related variables, pest incidence and water stress. The basic idea of site specific farming is not only to measure field variability, but also to be able to apply inputs at varying rates almost instantaneously, "real time", according to the needs. Variable rate application machinery is a type of field implements that could be used to handle field application of inputs such as seed, fertilizer and pesticides at the desired location in the field, at the right amount, at the right time and for the right reasons. The application of variable rate technology (VRT) can be accomplished either as a map based VRA or a sensor based VRA. However, different types of sensors are now available (or under development) that can monitor crop yield, soil properties, and crop condition that can be used to controlled field operations (Verma, et.al., 2020).

Management

Site specific farming makes farm planning both easier and more complex. The ability to combine information generated and the existing technology into a comprehensive and operational system is the third key area in precision farming. A farmer must adopt new level of management proficiency on the farm. Implicit in this is an increased level of knowledge of precision farming technologies such as GPS and GIS, better understanding of soil types, microclimates, aerial photography, economics of farming for accurate assessment of risk based on different decisions. The availability of yield map, weed distribution map, soil map, nutrient status map, water/ moisture availability map, pest and disease incidence map etc. require the farmer to make decision on how to treat the field for optimum or maximum yield. From this information, a treatment map or a DSS (Decision support system) can be developed utilizing GIS, agronomic, economic and environmental software, to help the farmer manage his field. The real values for a farmer is that he is enabled to easily and confidently manipulate seeding rate, plans more accurate crop protection and fertilizer application programmes, performs more timely tillage and knows the yield variation within a field. This precise micromanagement of his farming enterprise will enhance the overall cost effectiveness of precision farming in crop production. SSNM is likely to provide a greater profitability advantage for (a) high value crops, (b) areas where input cost are high and (c) areas where production conditions are very heterogeneous.

Fig. 4.5: Various component practices and technologies commonly associated with specific precision agriculture systems (IPNI, Reetz, Better Crops, 1994).

Managing the right source at the right rate, right time, and in the right place may be best accomplished with the right tools. Various technologies are available to help farmers and crop advisers make decisions related to nutrient management, from soil sampling to fertilizer application to yield measurement. These tools enhance the ability to fine-tune nutrient management decisions and develop the SSNM plan for each field. Farmers and the farm employees, management and agronomic advisers, and input suppliers are all part of a team, each contributing to the decision process in different ways.

Sustainable PA is this century's most valuable innovation in farm management that is based on using Information and Communication Technologies (ICTs). This is the most recent innovation technology based on sustainable agriculture and healthy food production and it consists of profitability and increasing production, economic efficiency and the reduction of side effects on the environment.

4.4 ROLE OF AGRICULTURAL EXTENSION

Similar to most developing countries, India is observing a structural transformation with the share of agriculture sector in total Gross Domestic Product (GDP) declining and that of non-agriculture (industry and services) increasing. Despite this decline, India continues to be predominantly an agrarian rural economy, with around 69 per cent of its population living in rural areas (Census 2011) and around 47% of the workforce engaged in agriculture (Labour Bureau, 2015-16). Given this huge dependency of rural households on agriculture, it has become imperative to focus on its growth in order to ensure food security and eliminate poverty in the country.

While agriculture growth depends on various factors ranging from rainfall at one end to investments in irrigation, agri-R&D, and prices on other, one of the critical factors is agriculture extension. It is this agri-extension that ensures that innovations in the labs are translated and implemented on the lands of peasants. In developing countries such as India, spending on agriculture is one of the most important government instruments for promoting economic growth and alleviating poverty in rural areas (Fan and Saurkar, 2006). Amongst various types of government spending for agriculture, agricultural R&E (including extension) is said to be one of the most critical for promoting farm yields (Fan, *et. al.*, 2008). In addition, several studies have shown the association of high profitability of agricultural research investment on agriculture production. Similar results are found in the upcoming study (Gulati, *et al.*, 2018) on the impact of public investment and subsidies on poverty alleviation and agricultural growth.

AGRICULTURAL EXTENSION

Extension is a dynamic concept in the sense that the interpretation of it is always changing. Extension, therefore, is not a term which can be precisely defined, but one which describes a continual and changing process in rural areas.

- Extension can be considered as an informal educational process directed toward the rural population. This process offers advice and information to help farmers solve their problems. Extension also aims to increase the efficiency of the family farm, increase production and generally increase the standard of living of the farm family.
- The objective of extension is to change farmers' outlook toward their difficulties. Extension is concerned not just with physical and economic achievements but also with the development of the rural people themselves. Extension agents, therefore, discuss matters with the rural people, help them to gain a clearer insight into their problems and also to decide how to overcome these problems. In agricultural-dependent economies, extension programmes have been the main conduit for disseminating information on farm technologies, support rural adult learning and assist farmers in developing their farm technical and managerial skills. It is expected that extension programmes will help increase farm productivity, farm revenue, reduce poverty and minimize food insecurity.

In India, public funding for agriculture R&E is contributed by both centre and state with around 55.4 per cent of the total allocation contributed by the centre and 44.6 per cent by states. However, as a percentage of gross domestic product from agriculture (GDPA), it amounts to about 0.54 per cent (2014-15). Moreover, spending on agriculture R&E shows that there are considerable variations across regions. Sectorwise break up shows that around 70 per cent of the total agriculture R&E budget is allocated to crop-husbandry itself, while only 10 per cent is allocated to animal husbandry and dairy development (Gulati et al., 2018)

AGRICULTURE EXTENSION SYSTEM: MAJOR PLAYERS IN INDIA

India has one of the largest agricultural research systems in the world. Currently, the public research system in India is led by the Indian

Council of Agricultural Research (ICAR), which has 5 multidisciplinary national institutes, 45 Central Research Institutes, 30 National Research Centres (NRCs), 4 bureaux, 10 project directorates, 80 All-India Co-Ordinated Research Projects (AICRPs)/networks and 16 other projects/ programmes. In addition, there are 29 State Agricultural Universities (SAUs) and one Central Agricultural University, which operates through 313 research stations. AICRPs are the main link between the ICAR and the SAUs. The ICAR has also Zonal Research Stations (ZRSs) and 200 sub-stations. The National Academy of Agricultural Research Management (NAARM) is another institution under ICAR to conduct research and training in agricultural research management. The ICAR has also established 8 Trainers' Training Centres (TTCs) and 611 Krishi Vigyan Kendras (KVKs) at the district level as innovative institutional models for assessment, refinement and transfer of modern agricultural technologies (Gulati et. al., 2018).

In addition, there are 23 general universities under the University Grants Commission (UGC), involved in agricultural research. Several scientific organizations such as the Council of Scientific and Industrial Research (CSIR), Bhabha Atomic Research Centre (BARC), National Remote Sensing Agency (NRSA), Ministries and government departments such as Ministry of Commerce, Department of Science and Technology, Department of Biotechnology, Department of Ocean Development, and more than 100 private and voluntary organizations and more than 105 scientific societies are involved in the agricultural R&E and 3 form the part of the national agriculture research system of India (Vision 2020, ICAR). This extensive agriculture research infrastructure not only conducts agriculture research but are also responsible for educating and providing extension services to the farmers.

AGRICULTURE EXTENSION SYSTEM: PUBLIC SECTOR

Currently, the agriculture R&E system in India is dominated by the public sector and is led by the Indian Council of Agriculture Research (ICAR). The Agricultural Technology Management Agency (ATMA) society registered at the district level was mandated to coordinate all on-going extension efforts in the district and converge and share resources in a targeted fashion. The National Mission on Agriculture Extension and Technology (NMAET) was initiated in order to take a holistic view of extension by making the system farmer-driven and increase accountability by restructuring and strengthening existing agriculture extension programmes to enable the delivery of technology and to improve the current agronomic practices of farmers (Gulati *et. al.*, 2018).

Other major players providing extension services in the public sector are Krishi Vigyan Kendras (KVK), State Agricultural Universities (SAU) and ICT-led extension interventions by Department of Agriculture Cooperation and Farmers Welfare (DACFW), Government of India. KVKs are field research units of the ICAR and are meant to test new seed varieties, agronomic practices, machinery etc. in field conditions across different agro-climatic zones before these are cleared for adoption by farmers. Additionally, they conduct farmer outreach programmes through on-farm demonstration plots, training etc.

The SAUs are another important arm for promoting extension activities in the states. While their main mandate is formal degree programmes in major agricultural disciplines, they provide extension and training support through the directorate of extension and education. The information flow is mainly from the universities to the KVKs which are responsible for training farmers (Gulati *et. al.*, 2018).

An important reform undertaken in recent years by the Ministry of Agriculture at the national level has been the increasing use of modern technologies and communication strategies to help educate farmers. Since ICT has significant potential to reach large numbers of farmers in a cost-effective manner several schemes have been initiated such as Farmer's Portal, m-Kisan, Kisan Call Centre, Kisan TV channel, Agriculture Clinic and Agriculture Business Centres, Agriculture Fairs and Exhibitions and community radio stations.

AGRICULTURE EXTENSION SYSTEM: PRIVATE SECTOR

Agriculture extension services by the private sector are mostly delivered by input dealers, such as those marketing seeds, fertilisers, pesticide and farm machinery. Fertiliser companies undertake extension activities by conducting farmer meetings, organizing crop seminars, arranging for soil testing facilities, adopting villages etc. Additionally, NGOs, such as Professional Assistance for Development Action (PRADAN), BAIF Development Research Foundation (earlier registered as Bharatiya Agro-Industries Federation) and Action for Food Production (AFPRO) are actively involved in promoting extension activities in more than one state. PRADAN has mainly focused on promoting livelihood for the poor in different sectors ranging from agriculture and natural resource management to micro-enterprise in rural areas across eight states in India. BAIF is also working on the development of livelihoods by engaging in livestock development, environment conservation, and water resource management across 16 states (Gulati *et. al.*, 2018).

There is a pressing need to address the lack of efficient extension services given to farmers in areas which are being plagued by over-production and low prices. Traditional approach to Extension needs to be changed and focused towards market-led extension. Farmer Producer Organization (FPOs) are plugging some of the weaknesses in existing extension system. Farmers who have diversified to high value agriculture for example floriculture are earning higher, but with higher risks. Major source of extension, training and monitoring of quality of produce is done by the FPO such as crop production, seeds and fertilizer knowledge, water management and use of modern machinery.

The services provided in the field on the advice of the private sector have produced good results and have been beneficial. In terms of sustainable and organic farming, farmers have tried to grow organic crops but there is still a hitch in accepting organic farming even though they hope to get a higher price for organic produce- no market, lower storage value and no trust in the market are some of the problems farmers face. Suggestions offered include first, setting up of Community Farm schools with useful demonstrations (giving knowledge on diversification towards other high value crops). Second, Community Skill Development centres to be able to absorb framers in forward and backward linkages with agriculture and not be dependent on one type of farming. Lastly, to educate farmers on and encourage community Enterprising ventures to capitalize on diverse markets (Gulati *et. al.*, 2018).

Rural farmers farming on small hectares of land can be attributed to conditions such as lack of adequate credit, lack of access to product market, lack of adequate extension contacts, among others. Among these constraints, inadequate extension services have been identified as one of the main limiting factors to the growth of the agricultural sector and rural community development at large. Thus, the role of agricultural extension today goes beyond the transfer of technology and improvement in productivity, but also, it includes improvement in farmers' managerial and technical skills through training, facilitation and coaching, among others.

Agricultural extension programmes include capacity building in good agricultural practices (GAPs), creating linkages among the value chain actors (input dealers, farmers, wholesalers and retailers) and other value addition techniques. Thus, wider dissemination of information regarding farmer skill development, the use of improved farm technologies, general farm management practices and easy access to input and output markets have been the fundamental principles underlying delivery of agricultural extension services. Agricultural extension is aimed primarily at improving the knowledge of farmers for rural development; as such, it has been recognized as a critical component for technology transfer. Thus, agricultural extension is a major component to facilitate development since it plays a starring role in agricultural and rural development efforts

Micronutrients- Impact On Human Health

5.1 MICRONUTRIENT DEFICIENCIES IN SOILS, PLANTS, ANIMALS AND HUMANS

Micronutrients play a key role in growth and development of plants, animals and humans. Role of micronutrients in growth and reproduction of plants, animals and humans is sufficiently documented. Essential micronutrients for plants are zinc (Zn), copper (Cu), iron (Fe), manganese (Mn), boron (B), molybdenum (Mo), chlorine (Cl) and nickel (Ni); for animals these are Zn, Fe, Mn, Cu, selenium (Se), iodine (I) and cobalt (Co) and for human Zn, Cu, Fe, Mn, Mo, Co, I, Se, F (fluoride), Cr (chromium) are essential. Boron is considered as beneficial trace element for animals and humans because it prevents losses of Ca and Mg from the body. Each micronutrient plays a specific role in plant, animal and human metabolism and their deficiency cannot be mitigated by substitution of other elements (Shukla, *et. al.*, 2018).

Soil

Micronutrients deficiency in soil has been assessed through various analytical techniques across many parts of the world. The deficiencies primarily occur due to the excessive use of fertilizers that high-yielding crop varieties demand, along with the lack of micronutrient supplementation.

In most of the Indian soil, micronutrients are present in sufficient amounts; however, their bio-availability for plant uptake is low; thus, soils in several Indian localities are inadequate. Boron deficiencies exist in an irregular pattern in Indian soils, ranging from 68% deficiency in red soils of Bihar to 2% in alluvial soils of Gujarat. Maximum B deficiency (54–86%) was observed in Alfisol soils of West Bengal and Assam, due to high rainfall that leads to a decrease in water-soluble B. Indo-Gangetic plains with saline soil exhibited a higher concentration of B, whereas the moderate level of B was recorded in Rajasthan and Madhya Pradesh. Copper deficient soil was observed in Kerala, Himalayan Tarai zone, Bihar, Uttar Pradesh and north Madhya Pradesh (Singh, 2001).

Despite the higher abundance of iron in the earth's crust, its plant-available concentrations are low in the alkaline soils of Indo-Gangetic regions. Soil analysis reports demonstrated that approximately 12% of Indian soils are Fe deficient. Manganese-deficient soil is rarely observed in India as only 1–5% of surface soil samples were found to be Mn deficient. Manganese deficiency was primarily observed in Haryana, Bihar, Punjab as well and Madhya Pradesh. Additionally, excess application of lime in red lateritic soils of Orissa resulted in Mn deficiency (Singh, et. al., 2006). Molybdenum deficiency has been observed 11% in total Indian soil including hill soils of Andhra Pradesh, Konkan and Malabar regions and north and north-eastern Himalaya regions (Singh, et. al., 2006, Bhupalraj et. al., 2002). On the contrary, calcareous alkaline soils of the Punjab region possess high available Mo contents and thus potential toxicity can be observed in crops. Out of 65,000 soil samples of India, 51.2% samples were found to be Zn deficient; thus, Indian soils are the most deficient in Zn in the world, with a widespread deficiency in Indo-Gangetic plains. The decline in Zn deficiency has been observed in the states such as Haryana, Punjab, Uttar Pradesh, Bihar, Andhra Pradesh and Madhya Pradesh (Dhaliwal, et. al., 2022).

Fertility of Indian soils is generally exacerbated with progressively emerging micronutrient deficiencies due to their catalysed removal under agricultural intensification. According to latest estimates, out of about 188.4 thousand tonnes (Tt) of micronutrients removed by 263 Mt of food grains produced, individual nutrient-wise removal is Zn - 23.9 Tt, Fe 110.6 Tt, Cu - 37.4 Tt, Mn - 63.3 Tt, B - 9.2 Tt and Mo - 0.99 Tt (Takkar and Shukla, 2015). Enhanced removal has resulted in 36.5, 12.8, 7.1, 4.2 and 23.4% of more than 2 lakh soil samples measuring deficient in Zn, Fe, Mn, Cu and B, respectively (Shukla and Behera, 2017).

Plants

Ideal concentrations of micronutrients in plants are 100, 100, 50, 20, 20, 6, 0.1 and 0.1 mg/kg of dry matter for Cl, Fe, Mn, B, Zn, Cu, Mo and Ni respectively. Plants show deficiency symptoms or enter into the hidden hunger condition when the concentrations of micronutrients fall below their respective critical concentrations (Table 5.1). Visual diagnosis of micronutrient disorders is a powerful tool for rapid identification of plant health linked to fertility, micronutrient availability, uptake and verification of soil or foliar test results. In general, there are four stages in development of micronutrients deficiency in plant:

Stage 1: Depletion of micronutrients stored in the body – diminishing degree of saturation of the carriers and enzymes,

Stage 2: Impairment of micronutrients dependent biochemical functions,

Stage 3: Measurable changes in cellular and physiological functions, and

Stage 4: Appearance of structural and functional lesions.

When a plant is deprived of particular nutrient, it reflects into impairment of biological and physiological functions (up to stage 3) before showing deficiency as lesions or clinical symptoms (stage 4). The first three stages are manifested in hidden hunger, which may cause significant loss in plant growth and development and ultimately reduction in yield, if not diagnosed through plant tissue analysis in time.

Table 5.1. Critical concentration of micronutrients in crop plants

Micronutrients	Crops	Critical concentration (mg/kg) for deficiency
Zn	Cereals	15
	Millets	15 – 20
	Legumes	7 – 20
	Vegetables (French bean)	36
	Oilseeds	12 - 25
B	Cereals	4 – 10
	Millets	7 – 15
	Legumes	3 – 15
	Vegetables	3 – 5
	Oilseeds	5 - 10
Mn	Cereals	25
	Millets	10
	Legumes	10-35
	Vegetables	30 – 40
	Oilseeds	5 - 18
Cu	Cereals	2 – 4
	Millets	2 – 3.5
	Legumes	4 – 8
	Vegetables	2 – 6
	Oilseeds	2 - 10

Indian Journal of Fertilisers, April 2018

Fig. 5.1: *Visual micronutrient deficiency symptoms of various crops*

Animals

Uptake of mineral nutrients by crops and pastures from soil is influenced by several factors such as soil types and their properties like soil acidity, soil moisture content, temperature, climatic condition, crop type and variety, crop management practices, application of fertilizer and organic matter and microbial activity of soil etc. Among the factors, soil acidity plays pivotal role in influencing trace mineral uptake by the crops and pastures. Higher soil pH leads to an enhanced biological availability of some trace elements such as Se and Mo, whereas with lower soil pH, availability of Se is less. But the uptake of some cationic trace elements like Cu is increased. Sometimes, the level of Cu, Zn, Mn, Fe and Co in crops is sufficient for optimum yields but is not adequate to meet the needs of livestock animals. For improving animal nutrition, state and zone-wise mineral nutrient deficiencies have been identified.

Humans

Micronutrient deficiencies are of growing public health concern. Estimated 40% of the world's population, mostly in low income countries is facing a problem of micronutrient malnutrition. While iron, iodine and vitamin A deficiency have long been endemic on a global scale, vitamin D, vitamin B12 and riboflavin deficiency are also of concern. Micronutrients enable the body to produce enzymes, hormones, and other substances that are essential for proper growth and development Widespread prevalent nutritional deficiencies of vitamin A, Fe, Zn, and iodine have been adversely influencing the health of women and young children. Shortage of micronutrients in the diet can limit growth, weaken immunity, cause xerophthalmia (an irreversible eye disorder leading to blindness), and increase mortality. Neutropenia and leucopoenia, skeletal defects and degradation of nervous system (Prasad *et. al.*, 1961), defective melanin synthesis, which manifests as depigmentation or hypopigmentation (lack of colour) of hair and skin, keratinization of hair, steely hair are other signs of micronutrient deficiencies in the humans (Davis and Mertz, 1987).

Recognition of the fact that there are relationships between soils and human health, date back to 1700 AD is referred in de Crevecoeur paper 'Letters from an American Farmer' which speak that *"Men are like plants; the goodness and flavour of the fruit proceeds from the peculiar soil and exposition in which they grow"*. McCarrison (1921) concluded that the fertility of a soil determined the vitamin content of food crops grown in it, and therefore it influenced the human health. He also speculated that soil bacteria could contribute to human diseases. Recognizing the importance of soil as the origin of many of the mineral elements necessary for human health, the USDA - Yearbook of Agriculture 1938 included three chapters on this aspect. Exposure to soil microorganisms is thought to be important in the prevention of allergies and other immunity related disorders (Rook, 2010). Soil is an important source of medicines, 78% of antibacterial agents approved between 1983 and 1994 had their origins in the soil (Pepper *et. al.*, 2009). Beyond antibiotics, approximately 40% of all prescription drugs have their origin in soil. Thus the idea that human health is linked with soil is not new but the scientific study on such aspect is a recent undertaking.

Micronutrients: Striding Towards Food and Nutrition Security

Agriculture faces a multifaceted challenge due to escalating population growth, degrading oil health, declining factor productivity, impending climate change etc. To feed the burgeoning population with nutritious and nourishing food, there is a need to improve both the quantity and quality of crop produce through optimal and efficient use of agricultural inputs. Of different inputs, micronutrients are of immense importance in ensuring agricultural sustainability and maintaining soil-plant-animal health.

Being emerged as one of the major constraints to crop productivity as well as quality, the issue of micronutrient deficiency in soils is highly related to food and nutrition security. In fact, micronutrient deficient soils lead to harvested produce with low content of micronutrients, which in long-run inflict their deficiencies in animal health too, thereby posing a threat toward agricultural sustainability vis-sa-vis food and nutrition security.

Seventeen elements in the Periodic Table are known to be required by all higher plants. Amongst these essential elements, nine are considered as macronutrients viz. carbon (C), hydrogen (H), oxygen (O), nitrogen (N), phosphorus (P), potassium (K), calcium (Ca), magnesium (Mg) and sulphur (S), which normally remain in the plant tissues at concentrations of greater than 0.1% (dry weight). Eight others are micronutrients viz. zinc (Zn), boron (B), molybdenum (Mo), chlorine (Cl), copper (Cu), iron (Fe), manganese (Mn) and nickel (Ni), which are absolutely required in relatively smaller quantities for completion of life cycle of the organisms (plants or humans).

Human nutrition is not only influenced by the presence of these micronutrients, there has an impact of certain other elements viz. selenium (Se), iodine (I), fluorine (F), chromium (Cr), vanadium (V), silicon (Si), arsenic (As) and cobalt (Co) - commonly cobalamine, being absorbed by plants from soil and water and moved to animal and human gut through food chain.

Deficiencies of micronutrients in soils and plants are attributed to intensive cropping practices under high-yielding varieties of crops with little or negative addition of micronutrient fertilisers, limited use of organic manures/crop residues, use of high analysis fertilisers etc. For example, Zn is reported to be the most deficient across the globe. B comes to the next in order. There are certain others including Mo, Cu, Mn and Fe. Indian soils are mainly deficient in Zn, followed by B, Fe, Mn and Cu. The most important micronutrients for the State of West Bengal (W.B.) are Zn, B and Mo. When the deficiency is marginal, the plant does not exhibit any visible signs, and the deficiency is not easy to diagnose. This is known as hidden hunger, which causes the plant to yield less than expected. Even the nutritional health of soil affects the health of humans and livestock, causing malnutrition. The malnutrition due to micronutrients, especially Zn and Fe, is cropping up at a faster rate. Biofortification (agronomic/ genetic) appears to be a promising tool to combat micronutrient deficiency-cum-malnutrition.

A recent study through on-station and on -farm experiments across different districts in W.B. revealed that the deficiency alleviation of Zn and B as well as productivity improvement of transplanted rice can be done through (i) nutri-priming of seeds with 2.0% zinc sulphate heptahydrate ($ZnSO_4.7H_2O$, 21.0% Zn) solution for 25-30 hours before sowing, (ii) basal application of 2.5 g $ZnSO_4.7H_2O$ + 1.0 g borax or sodium tetraborate decahydrate ($Na_2B_4O_7.10H_2O$, 10.5% B) per sqm. or foliar spray of 0.2% $ZnSO_4.7H_2O$ + 0.1% disodium octaborate tetrahydrate ($Na_2B_8O_{13}.4H_2O$, 20.0% B) + 1.5% urea solution at the seedling age of 15 days in nursery, and (iii) two rounds of foliar spray with 0.2% $ZnSO_4.7H_2O$ + 0.1% $Na_2B_8O_{13}.4H_2O$ + 1.5% urea solution at maximum tillering and before panicle initiation in main field (spray volume of 300 and 600 L ha-1 at first and second spray, respectively). In case of Mo deficiency, 0.02% ammonium molybdate $[(NH_4)_6Mo_7O_{24}.4H_2O$, 52.0% Mo] solution can be added in the spray solution as scheduled above for second spray in main field.

Micronutrient balance in the soil-cropping system depends upon removal or depletion of micronutrients from soils by different crops and their replenishment over the years through fertilisers, manures, irrigation, rainwater etc. Required in small quantities, these are as essential as macronutrients, because one million atoms of N are of no use if one Mo atom is not present. Thus, micronutrients play a macro role in enhancing the use efficiency of NPK, augmenting crop yields, and addressing the problem of malnutrition in human beings and livestock as well. Hence, there is a need to develop site-specific nutrient management practices, utilizing the native (soil) and applied (straight/ chelated/fortified micronutrients, multinutrient mixtures, customised fertilisers, fritted/ slow release fertilisers, nano-fertilisers bio-release smart fertilisers and/or manures) sources of micronutrients in a more judicious and efficient way, in balanced amount and right proportion for achieving sustainable agricultural productivity, besides maintaining nutritional quality of food/ feed and keeping environmental pollution under check. Diversification of the cropping system and balanced fertilisation for meeting nutritional requirements is essential for alleviating malnutrition and achieving socio-economic well-being.

Each and every farmer should make use of the Soil Health Card in understanding the deficiency status of the soil and applying nutrients *as per prescription and application schedule. Many farmers often tend to mix different spray solutions to save cost and time by applying them together. Care should be taken to see the compatibility, use the same as soon as possible after mixing if done, and not to mix them with phosphate based solutions as they may get fixed and become unavailable to the plants. Farmers are to be more vigilant while purchasing micronutrient products, and should buy the products from reputed dealers and not hesitate in clarifying doubts to procure the right product that is right for their soil.*

To meet the escalating demand of nutritious food/feed/fodder for growing human and livestock population, while reducing agriculture's footprint on the environment, it is imperative to have a balanced fertilisation, including an integrated micronutrient management schedule in order to stride over the issue of micronutrient deficiency cum-malnutrition. Accomplishing food and nutrition security with respect to micronutrients is a challenge for the societal welfare. Finally, what we need is the sustenance of soil health with the appropriate replenishment of micronutrients as because healthy soils will produce healthy food and ensure healthy life.

- Sampad R. Patra
Department of Agriculture (Govt. of W.B.),
(Director of Agriculture & Ex-Officio Secretary, W.B.)
IMMA Souvenir, 2018

5.2 MICRONUTRIENT-RELATED MALNUTRITION

Micronutrients are vitamins and minerals needed by the body in very small amounts. However, their impact on a body's health are critical, and deficiency in any of them can cause severe and even life-threatening conditions. They perform a range of functions, including enabling the body to produce enzymes, hormones and other substances needed for normal growth and development. Micronutrient deficiencies also referred to as 'Hidden Hunger' affects the health, learning ability as well as productivity owing to high rates of illness and disability contributing to vicious cycle of malnutrition, underdevelopment and poverty.

Deficiencies in iron, vitamin A and iodine are the most common around the world, particularly in children and pregnant women. It is estimated that around two billion people in the world are deficient in one or more micronutrients. Micronutrient deficiencies (such as iodine, iron and vitamin A deficiency) not only affect the health but are also projected to cost around 0.8-2.5 per cent of the gross domestic product (Stein and Qaem, 2007). Low- and middle-income counties bear the disproportionate burden of micronutrient deficiencies. In India, around 0.5 per cent of total deaths in 2016 were contributed by nutritional deficiencies (Disease study Initiative).

Inadequacies in intake of vitamins and minerals often referred to as micronutrients, can also be grouped together. Iodine, vitamin A, and iron are the most important in global public health terms; their deficiency represents a major threat to the health and development of populations worldwide, particularly children and pregnant women in low-income countries. (WHO, 2021)

From a public health viewpoint, Micronutrient Malnutrition (MNM) is a concern not just because such large numbers of people are affected, but also because MNM, being a risk factor for many diseases, can contribute to high rates of morbidity and even mortality. It has been estimated that micronutrient deficiencies account for about 7.3% of the global burden of disease, with iron and vitamin A deficiency ranking among the 15 leading causes of the global disease burden (WHO Report, 2002). According to WHO mortality data, around 0.8 million deaths (1.5% of the total) can be attributed to iron deficiency each year, and a similar number to vitamin A deficiency.

VITAMIN DEFICIENCIES

Vitamin A deficiency is the leading cause of preventable blindness in low- to middle-income nations, with children and pregnant women being particularly susceptible.

The WHO estimate that 25 % of the population are vitamin D deficient worldwide. Currently, Public Health England advise that a daily supplement containing 10 μg vitamin D is taken during the autumn and winter months. During these months, sunlight is of insufficient intensity for vitamin D3 synthesis, thus reliance on dietary sources of vitamin D to meet requirements is increased. It is difficult to achieve this level of intake from natural food sources alone, thus alternative approaches are needed.

Emerging evidence suggests that low vitamin B12 status is associated with diseases of ageing including cognitive dysfunction, CVD and osteoporosis (Hetzel and Pandav,1994). As older people are particularly susceptible to sub-clinical deficiency, due to food-bound B12 malabsorption, this is important from a public health perspective, especially considering our ageing population.

Folates, a group of water-soluble B vitamins, have a predominant role in primary metabolism; thus, detrimental physiological effects occur upon folate deficiency. Enzymes utilize folate in thymidylate, purine synthesis and pantothenate (VitB5) formation. The critical role of folate in the prevention of neural tube defects is widely acknowledged and in countries with mandatory folic acid fortification, substantial reductions in neural tube defects have been observed.

MINERAL DEFICIENCIES

Iron deficiency anaemia (IDA) is the most common nutritional disorder in the world, affecting almost two billion people, particularly pregnant women and children in developing countries. Iron deficiency is a leading cause of anaemia which is defined as low haemoglobin concentration. Anaemia affects 40% of children younger than 5 years of age and 30% of pregnant women globally. Anaemia during pregnancy increases the risk of death for the mother and low birth weight for the infant. Worldwide, maternal and neonatal deaths total between 2.5 million and 3.4 million each year (Stevens *et. al.*, 2013)

Fig. 5.2: Effects of micronutrient malnutrition on human health.

Globally, 17.3% of the population is at risk for zinc deficiency due to dietary inadequacy; up to 30% of people are at risk in some regions of the world (Wessells and Brown, 2012). Providing zinc supplements reduces the incidence of premature birth, decreases childhood diarrhoea and respiratory infections, lowers the number of deaths from all causes, and increases growth and weight gain among infants and young children. Providing zinc supplementation to children younger than 5 years appears to be a highly cost-effective intervention in low- and middle-income countries (Wessells and Brown, 2012).

Iodine is a key nutrient for brain development. Although severe deficiency is now relatively rare due to iodised salt programmes within countries, mild-to-moderate iodine deficiency during pregnancy is a public health problem in many countries, including the UK. Mild-to-moderate deficiency has been associated with impaired neurodevelopment of the child (Thilly *et. al.*, 1980).

Table 5.2 Prevalence of the three major micronutrient deficiencies by WHO region

WHO region	Anaemia [a] (Total population)		Insufficient [b] iodine intake (Total population)		Vitamin A [c] Deficiency (Pre-school children)	
	No (millions)	% of total	No (millions)	% of total	No (millions)	% of total
Africa	244	46	260	43	53	49
Americas	141	19	75	10	16	20
South-East Asia	779	57	624	40	127	69
Europe	84	10	436	57	No data available	
Eastern Mediterranean	184	45	229	54	16	22
Western Pacific	598	38	365	24	42	27
Total	2030	37	1989	35	254	42

Sources: WHO, 1995, 2001 and 2004
a Based on the proportion of the population with haemoglobin concentrations below established cut-off levels.
b Based on the proportion of the population with urinary iodine <100µg/l.
c Based on the proportion of the population with clinical eye signs and/or serum retinol ≤0.70µmol/l.

Many of these deficiencies are preventable through nutrition education and consumption of a healthy diet containing diverse foods, as well as food fortification and supplementation, where needed. These programmes have made great strides in reducing micronutrient deficiencies in recent decades but more efforts are needed.

Interest in micronutrient malnutrition has increased greatly over the last few years. One of the main reasons for the increased interest is the realization that micronutrient malnutrition contributes substantially to the global burden of disease. In 2000, the World Health Report1 identified iodine, iron, vitamin A and zinc deficiencies as being among the world's most serious health risk factors. In addition to the more obvious clinical manifestations, micronutrient malnutrition is responsible for a wide range of non-specific physiological impairments, leading to reduced resistance to infections, metabolic disorders, and delayed or impaired physical and psychomotor development. The public health implications of micronutrient malnutrition are potentially huge, and are especially significant when it comes to designing strategies for the prevention and control of diseases such as HIV/AIDS, malaria and tuberculosis, and diet-related chronic diseases.

Another reason for the increased attention to the problem of micronutrient malnutrition is that, contrary to previous thinking, it is not uniquely the concern of poor countries. While micronutrient deficiencies are certainly more frequent and severe among disadvantaged populations, they do represent a public health problem in some industrialized countries like iodine deficiency in Europe, where it was generally assumed to have been eradicated, and of iron deficiency, which is currently the most prevalent micronutrient deficiency in the world. In addition, the increased consumption in industrialized countries (and increasingly in those in social and economic transition) of highly-processed energy-dense but micronutrient-poor foods, is likely to adversely affect micronutrient intake and status.

Table 5.3 Selected micronutrient deficiencies, their risk factors and effects

Micronutrient	Prevalence of deficiency	Risk factors	Health consequences
Iron	2 billion cases of anaemia worldwide. There are approximately 1 billion cases of iron-deficiency anaemia and a further 1 billion cases of iron deficiency without anaemia worldwide	• Low intakes of meat/fish/poultry • High intakes of cereals and legumes • Pre term delivery / Low birth weight • Heavy menstrual losses • Infections and diseases	• Reduced cognitive performance • Lower work performance and endurance • Impaired iodine and vitamin A metabolism • Anaemia • Increased risk of maternal mortality and child mortality (with more severe anaemia)

Zinc	Insufficient data, but prevalence of deficiency is likely to be moderate to high in developing countries especially those in Africa, South-East Asia and the Western Pacific	• Low intakes of animal products • High phytate intakes • Malabsorption and infection	• Possibly poor pregnancy outcomes • Impaired growth (stunting) • Decreased resistance to infectious diseases • Severe deficiency results in dermatitis, retarded growth, diarrhoea, mental disturbance, delayed sexual maturation and/or recurrent infections
Iodine	An estimated 2 billion people have inadequate iodine nutrition and therefore are at risk of iodine deficiency disorders	• Residence in areas with low levels of iodine in soil and water • Living in high altitude regions, river plains or far from the sea	• Birth defects / Still birth • Cognitive and neurological impairment including cretinism • Hypothyroidism • Goitre
Vitamin A	An estimated 254 million preschool children are vitamin A deficient	• Low intakes of dairy products, eggs and β-carotene from fruits and vegetables	• Increased risk of mortality in children and pregnant women • Night blindness, xerophthalmia
Vitamin B1 (thiamine)	Insufficient data on marginal deficiency	• High consumption of refined rice and cereals • Low intakes of animal and dairy products, and legumes • Chronic alcoholism	• Beriberi • Wernicke-Korsakov syndrome (usually in alcoholics) with confusion, lack of coordination and paralysis
Vitamin B2 (riboflavin)	Insufficient data,	• Low intakes of animal and dairy products • Chronic alcoholism	• Fatigue, eye changes, • Dermatitis • Brain dysfunction • Microcytic anaemia • Impaired iron absorption and utilization

Vitamin B$_3$ (niacin)	Severe deficiency (pellagra) still common in Africa, China and India and recently reported among displaced populations (south-eastern Africa) and in famine situations	• Low intakes of animal and dairy products • High consumption of refined cereals	• Pellagra, • Dermatitis, • Diarrhoea, vomiting, • Neurological symptoms, depression and loss of memory
Vitamin B$_6$	Insufficient data, but recent reports suggest that deficiency is likely to be widespread in developing countries	• Low intakes of animal products • High consumption of refined cereals • Chronic alcoholism	• Neurological disorders and convulsions, • Dermatitis, • Anaemia
Folate (vitamin B$_9$)	Insufficient data	• Low intakes of fruits and vegetables, legumes and dairy products	• Megaloblastic anaemia
Vitamin C (ascorbic acid)	Severe deficiency (scurvy) regularly reported in famine situations (e.g. East Africa)	• Low intakes of fresh vitamin C-rich fruits and vegetables • Prolonged cooking	• Severe deficiency results in scurvy with haemorrhagic syndrome (i.e. bleeding gums, joint and muscle pain, peripheral oedema) • Anaemia
Vitamin D	Higher at more northerly and southerly latitudes where daylight hours are limited during the winter months	• Low exposure to ultra-violet radiation from the sun • Wearing excess clothing • Having dark pigmented skin	• Severe forms result in rickets in children and • Osteomalacia in adults

Sources: WHO- Geneva, 1995, 1996, 2001 and 2004

In the wealthier countries, higher incomes, greater access to a wider variety of micronutrient-rich and fortified foods, and better health services, are all factors that contribute to the lowering of the risk and prevalence of MNM. However, consumption of a diet that contains a high proportion of energy-dense but micronutrient-poor processed foods can put some population groups at risk of MNM. Although at present this practice is more common in industrialized countries, it is rapidly becoming more prevalent among countries undergoing social and economic transition. Table 5.3 provides an overview of the prevalence, risk factors, and health consequences of deficiencies in each of the 15 micronutrients covered in these guidelines.

Estimated 40% of the world's population, mostly in low income countries is facing a problem of micronutrient malnutrition. Burgeoning rise in the number of the people affected with micronutrient malnutrition during last four decades in India coincides with expansion of the area under rice-wheat or rice-rice cropping systems (low nutritional quality food) having high yielding varieties (HYVs) at the expense of traditional cereal-pulses/legume (high nutritional quality food) systems. In South Asia, with the introduction of HYVs a spectacular increase in production of wheat (400%) and rice (200%) over four decades is associated with increasing incidences of Fe deficiency anaemia among nonpregnant and pre-menopausal women caused by decrease in Fe density (mg Fe/ kcal of available food) in diet. Consequences of micronutrient deficiencies have grave implications on health, livelihood and well-being of the afflicted people (Combs, *et al.*, 1996). Dietary intake ought to be changed with increased consumption of pulses to ensure adequate and balanced micronutrient supply to one and all in an affordable manner (Welch *et. al.*, 1997).

5.3 IMPACT OF MICRONUTRIENT DEFICIENCIES AND TOXICITIES ON CROP PRODUCTIVITY AND ANIMAL AND HUMAN HEALTH

Agriculture is the primary supplier of food and nutrients to all human on earth. If agriculture cannot provide adequate amounts of all nutrients, the food systems become dysfunctional and malnutrition arises. The biggest challenge is to feed and nourish a burgeoning human population with limited land resources for productive agriculture. In addition, multimicronutrient deficiencies in soils worldwide leading to production of poor quality crop produce, particularly low in trace elements, ultimately affect the animal and human health (Sillanpaa, 1990; Shukla *et. al*, 2014, 2016; Behera and Shukla, 2014). Toxicity of trace elements (particularly Fe, Al, Se, As, F, Cr etc.) in soil and water may also affect animal and human health (Shukla *et. al.*, 2018).

Zinc

The available Zn in Indian soils ranges from 0.01 to 52.9 mg/kg. Acute Zn-deficient soils are intensively cultivated ones characterized by coarse texture (sandy/ loamy sand), high pH (> 8.5 or alkali/ sodic soils) and/or calcareousness, and low soil organic carbon content (< 0.4%) (Shukla, *et. al.*, 2014). Zinc deficiency witnessed a decline from 46% in 1967-1987 to 36.5% in 2011-2017 due to regular and higher use of Zn fertilizer in some parts of the country (Shukla, *et. al.*, 2018). Extensive research and extension activities on micronutrients led to an increase in Zn fertilizer use linearly. Zn deficiency decreased to 36.5% in 2017 and based on the current trends, Zn deficiency would reduce to 21% by 2025-30.

Zinc deficiency disorders are known by different nomenclatures like khaira disease in rice, rosetting in wheat, white bud in maize, little leaves and mottling in vegetables, and reduced fruit formation in citrus. Out of 4,144 trials conducted on farmers' fields during 1967-84, 58% exhibited response to Zn application (Takkar *et. al.*, 1989; Singh, 2001). Number of trials responding to applied Zn increased over the years from 58% during 1967-1984 to 63% during 1985-2000, 72% during 2000-2010, to 80% during 2011-2016. This indicates that either new cultivars are more responsive to Zn application or its deficiency has intensified due to greater mining of Zn from soil without its matching replenishment.

Feed and fodders produced on micronutrient-deficient soils and fed to cattle in Haryana resulted in higher percentage of Zn, Cu and Mn deficiency in their blood serum, hair and milk. Survey in Vadodara district of Gujarat showed that dry fodder tested low in Fe (61%), Zn (72%) and Cu (87%) and green fodders were low in Fe (17%), Zn (5%) and Cu (23%) although most of the soils on which this was grown were adequate in terms of available Fe, Mn and Cu (Shukla *et. al.*, 2018).

A specific disorder resulting from zinc deficiency in animals is parakeratosis – a disorder of the epidermal layer of the skin occurring in calves, sheep, goats and piglets. Phytate (inositol hexaphosphate): Zn ratio plays an important role in influencing the Zn bioavailability in animal body. Excess intake of Zn is relatively rare in farm animals. However, excess zinc may reduce the digestibility of phosphorus, and cause anemia and digestive disorders. Poisoning is conditioned primarily by the antagonistic relationship of zinc with iron and copper. Excessive intake of Zn additives may lead to the essential fatty acid metabolism which influences synthesis of prostaglandin.

In humans, Zn deficiency was recognized as a health concern for the first time in 1961. Prasad *et. al.* (1961, 1963) described the first cases of human Zn deficiency syndromes: growth stunting, delayed sexual development and hypogonadism in young adults from Iran and Egypt. Besides, Zn deficiency leads to diarrhoea, respiratory malfunctions, weak immune system, impaired cognitive function, neuronal atrophy, behavioural problems, memory impairment, spatial learning, lesions on dermal tissue/ keratin and parakeratosis.

It is estimated that one-third of the world population living in developing countries suffers from high risk of zinc deficiency. The vulnerable populations include infants, young children, and pregnant and lactating women because of their higher zinc requirements at critical stages of growth and physiological needs. In general, a very strong correlation has been reported between soil Zn status and human Zn deficiency level (Singh, 2009; Shukla *et. al.*, 2016). In India, about 25% of the total population suffers from Zn deficiency. The prevalence of nutritional stunting due to Zn deficiency in India is about 47.9% in children of less than 5-year age against average of 33% in the world's population.

Iron

In India, the problem of iron deficiency mainly occurs in calcareous and alkaline soils with pH > 7.5. Availability of Fe gets aggravated under drought or moisture stress conditions due to conversion of ferrous form of iron (Fe^{2+}) into less available ferric form (Fe^{3+}). Sometimes, high concentrations of P, $NO^{3-}N$ and organic matter contents also hinder iron availability to the crop plants. Strongly acid and waterlogged soils have very high level of available Fe. There is a peculiar problem in flooded (paddy) rice soils where rice yields get severely reduced by Fe toxicity. The soils of north-east region, Odisha and Kerala suffer from Fe toxicity problem in rice paddies (Shukla *et. al.*, 2018).

Iron chlorosis in plants, also called lime-induced chlorosis, is generally observed in upland crops especially aerobic rice, sorghum, groundnut, sugarcane, chick pea grown in Fe-deficient highly calcareous soils, compact soil with restricted aeration, soils with low in active Fe and high in P and bicarbonate content.

Iron is one of the most abundant element (4th at 5%) on Earth. Yet its deficiency is probably the most common across the world affecting as many about 2 billion people (over 30% of the world's population). Globally, Fe-deficiency anaemia is the most common widespread nutritional disorder, affecting more than 50% pregnant women and 40% of infants and preschool children. Although only 14.4% Indian soils are deficient in available Fe, iron deficiency anaemia (IDA) is quite acute and widespread in marginalised section of our country.

In some pockets, a strong relationship exists between available Fe level in soils and occurrence of IDA. The low Fe content in forage and food grains produced in arid and semi-arid regions is attributed to acute Fe deficiency in these soils (34% in Rajasthan, 26% in Gujarat, 22% in Haryana, and 23% in Maharashtra). However, in some humid and sub humid regions of the country where Fe is in sufficient amount in soil, increased IDA may be attributed to poor accumulation of Fe in plants grown on these soils.

Copper

Copper deficiency in Indian soils is almost negligible. Available Cu content in Indian soils ranges from 0.01 to 136.4 mg/kg with an average value of 2.05 mg Cu/ kg (Shukla and Tiwari, 2016). Copper availability is mainly influenced by pH, SOC, $CaCO_3$ and clay content in soils. It increases with increase in organic matter and clay content, while decreases with increase in pH and $CaCO_3$ content of the soil (Katyal and Agarwal, 1982; Rattan et. al., 1999).

Crops grown on severe Cu-deficient soils results in reduced yields and poor crop quality. These include shrivelled grains and reduced viability of seeds in cereals. In citrus, abnormal shaped fruits with a rough exterior, low juice content and poor flavour and in apples, small fruits of poor quality are found due to Cu deficiency.

Cu deficiency is responsible for leukoderma (Vitiligo), depigmentation of hair and skin around the brisket, neck, face, hind limbs and abdomen in buffaloes (Randhawa, 1999; Sinha et. al., 1976). Copper deficiency also caused 'falling disease' in milch cows. Copper concentrations in the livers of affected animals was lower (32.6 mg kg^{-1}) as compared to healthy cow (55.7 mg kg^{-1}) (Vasudevan, 1987). Post parturient haemoglobinuria (PPH), molybdenosis induced hypocupraemia, commonly referred as nutritional haemoglobinuria, was observed, in high yielding cattle and buffaloes in India (Dhillon and Dhillon, 1991; Singh and Randhawa, 1990). Molybdenosis in buffaloes in Punjab and hypophosphatemia in Maharashtra have also been reported (Dhillon and Dhillon, 1991).

Copper deficiency in human leads to hypocupraemia, neutropenia and leucopoenia, degradation in nervous system, skeletal defects, hypopigmentation of hair and skin, keratinization of hair, cardiovascular disorders, osteoporosis, arthritis and infertility (Davis and Mertz, 1987).

Manganese

The disease prone areas of Maharashtra had high levels of Ca and Mo and very low level of P and Cu in the feeds and fodder (Hassan *et. al.*, 1985).

The available Mn content varies from 0.01 to 445.0 mg/kg with a mean value of 21.8 mg/kg (Shukla *et. al.*, 2014). In general, Mn deficiency problems occur on soils with low total contents of Mn (heavily weathered tropical and sandy soils), on peaty soils, or organic-rich soils with a pH above 6, and on mineral soils with pH values of 6.5 or above, calcareous soils, or acid soils which have been heavily limed. Similar to Fe, incidence and severity of Mn deficiency accentuates in crops grown at very low moisture content. On the other hand, Mn is more mobile in imperfectly drained soils (waterlogged) and often Mn toxicity is observed in rice grown under continuous submerged conditions on such soils.

Deficiency of Mn results in appearance of greenish-grey specks at the lower base of younger leaves in monocots, which finally become yellowish to yellow-orange. It may lead to development of marsh spots (necrotic areas) on the cotyledon of legumes. In sugarcane, Mn deficiency is named as Pahala Blight.

Crops like wheat grown in Mn deficient soils or hidden hunger of Mn in Haryana not only produced low yields but led to infertility in cattle due to low Mn content in fodder and grain. Evidence of increased infertility was recorded in cattle fed with low Mn fodder grown in highly calcareous soils (free $CaCO_3$ content 20-48%) in Pusa, Bihar (Singh, 2009). Productivity of these animals was low and their blood serum Mn concentration was also low as compared to that in cattle fed with fodders grown in Mn-adequate soils.

Boron

In India, extent of B deficiency is next only to Zn. Total B content in Indian soils ranges from 2.6 to 630 mg/kg (Takkar, 2011) and available (hot water soluble – HWS) B ranges from 0.04 to 250 mg B kg-1, with an average of 21.9 mg / kg soil (Shukla and Tiwari, 2016). Availability of B to plants is governed by soil pH, $CaCO_3$ and organic matter contents. In addition, total B content in soil, its interactions with other nutrients, plant type or variety and environmental factors also have strong influence on B availability. By and large, B deficiency adversely impacts the crop productivity in highly calcareous soils, sandy leached soils, limed acid soils or reclaimed yellow or lateritic soils.

Boron deficiency symptoms first appear on the growing tips and younger leaves with stunted plant growth. It results in production of hollow heart in peanut, black heart in beet, distorted and lumpy fruit in papaya, and hollow pith in cabbage and cauliflower.

Boron is a newer trace element identified for nutritional role in animals and its biological importance and dietary essentiality is unclear. Boron supplementation in farm animals enhanced the immunity by increasing the serum levels of tumour necrosis factor and interferon-gamma (Shukla et. al., 2018). The dietary B prevents metabolic disorders in peri-parturient dairy cows by increasing the serum Ca levels under a situation of high physiological demand.

Since B works with Mg, it has been suggested that B can be beneficial to persons suffering from hyperthyroidism or persons with thyroid disease who have low magnesium symptoms like rapid heart rate and muscle cramping. Boron lessens the effects of low Mg in diet on body growth, serum cholesterol and ash concentration in bone. Boron supplementation increases estrogen and testosterone levels and helps in insulin and glucose metabolism (Hunt, 2004). Higher intake of B can cause toxicity which manifests itself in nausea, vomiting, diarrhoea, dermatitis, lethargy, and increased urinary excretion of riboflavin (Ziegler and Filer, 1996).

Molybdenum

Molybdenum is least studied micronutrients in India. Total Mo in Indian soils ranges between 0.1 to 12 mg/kg and available Mo, extracted with ammonium oxalate (pH 3.3), varies from traces to 2.8 mg / kg (Behera *et. al.*, 2011, 2014). Most of the soils are adequate in Mo but its deficiency is noticed in some acidic, sandy and leached soils. Molybdenum deficiencies severely affect legumes, crucifer vegetables and oilseed crops on acid and severely leached soils.

Molybdenum deficiency results in stunted plant growth and restricted flower formation. It causes whiptail disease of cauliflower.

Molybdenum deficiency is reported in eastern high rainfall zone soils, low in available Mo. In northern parts of West Bengal, problem of hair and hooves falling in cattle has been reported widely due to low Mo in alluvial leached soils. Grain legumes have higher concentration of Mo. Toxicity of Mo in animals and human is reported in some parts of Punjab which affects Cu utilization in the body due to interaction of Mo with Cu (Nayyar *et. al.*, 1990).

In human, Mo helps in breakdown of build-up of toxic compounds such as sulphites. Molybdenum has been shown to help fight cancer-causing nitrosamines and even assists in preventing cavities. The most commonly associated function of molybdenum is its role in the production of uric acid. Adequate Mo prevents dental caries, mouth and gum disorders, oesophageal cancer, and sexual impotence in old people (Shukla *et. al.*, 2018).

Nickel

Total Ni content in soils of India ranges between 20-1000 mg kg-1whereas available status ranges between 0.2-0.8 mg kg^{-1} soils (Singh, 2009). Its deficiency in Indians soils has not yet been reported. It is constituent of urease enzyme, required for the breakdown of urea to liberate the nitrogen into a usable form for plants. Seeds need nickel in order to germinate and grow, and also for absorption of iron.

Nickel deficiency in plants is linked to production of dwarf foliage and reddish pigmentation in young leaves.

In humans, Ni is important element for the heart muscle, liver and the kidneys. It is involved in hormone, lipid and membrane metabolism.

Cobalt

Total Co content in Indian soils varies widely from traces to 277 mg/kg and depends largely on the parent rock and climate of the region. Available Co content ranges from 0.06 to 2.1 mg/kg soil and is markedly influenced by factors such as texture, pH, $CaCO_3$, organic matter content, soil-crop management systems and practices.

Cobalt deficiency in soil leads to its deficiency in plants and affects animals' productivity. Availability of Co to plant is poor on soils high in MnO_2 content because of higher affinity of MnO_2 to Co (Takkar, 2015).

Cobalt helps in nodulation and growth of legume and brassica crops. Seed treatment with cobalt containing salts @ 2 to 3 g kg-1 seed is helpful in increasing the seed yield. Cobalt is essential for synthesis of vitamin B12, mainly in ruminant cattle.

Vanadium

Vanadium content in soils and waters is primarily determined by the geological parent material (Hope, 1997). Vanadium inputs to soils also come from human activities such as addition of phosphate fertilizers, soil amendments and road-fill materials derived from steel slag (Molina et. al., 2009). Vanadium is absorbed by plant as vanadate ion (Ullrich-Eberius et. al., 1989) and can inhibit the plasma membrane hydrogen (H+) translocation ATPase, which is known to play an important role in nutrient element uptake by plant cells.

Vanadium is required in very small quantity by human. It is also rapidly utilized and excreted through urine. Inadequate V leads to lower birth weight of babies. Higher V levels in soils may cause human health problems like distal renal tubular acidosis (Tosukhowong et. al., 1999).

Selenium

Selenium is a trace element not required by most of the cultivated crops, yet maximum crop yield is obtained on soils characterized by traces of Se. In soils, Se is found in selenate and selenite forms, the latter being more soluble and available to plants than the former

Livestock fed with the low Se-feed might suffer from serious muscular disorders and other diseases. White muscle disease due to Se deficiency is probably the most common and serious disorder found in calves and lambs. Cases of Se toxicity have been reported from certain parts of Punjab in Pakistan and northwest states of India (Takkar, 1996). Diseases caused by Se toxicity is named as 'Alkali disease' in the USA and 'Degnala' in India and Pakistan. Though chronic form of syndromes appear throughout the year, but it is more prevalent in the post-rainy season. The animals develop lesion on their tail, ear tips and limps. Skin and hooves are the worst affected. In very severe cases, the hooves fall off and the animals ultimately succumb. (Shukla *et. al.*, 2018)

Selenium content in food crops and water is important to human health and on this strength it is now recognized as a beneficial element. Importance of Se to human health (in terms of antioxidant, anti-inflammatory, anti-cancer, anti-viral, and antiaging activity, along with key roles in the thyroid, brain, heart, and gonads) is highlighted by its status as the only micro-nutrient to be specified in the human genome, as selenocysteine, the twenty-first amino acid (Shukla *et. al.*, 2018)

Iodine

Iodine is not essential for plant growth and development, but lack of iodine in the soil results in plants poor in iodine. Iodine deficiency is more prevalent in soils and waters of hilly and mountainous areas due to continuous washout of iodine by rain water from the surface soil. Iodine level in the drinking water indicates its content in soil. Iodine is loosely held in soil and thus washed off to sea with rainwater

Iodine is one of seven generally recognized micro minerals needed in the diets of dairy cattle and other animals. It is unique among minerals, because a deficiency leads to a specific and easily recognizable thyroid gland enlargement, called goitre. Iodine is important in the synthesis of the thyroid hormones, thyroxine (T4) and triiodothyronine (T3), that regulate energy metabolism in animals.

In humans, iodine is a principal component of thyroid hormones, thyroxine (T4) and triiodothyronine (T3), which are essential for normal growth, physical and mental development in humans. Iodine deficiency affects brain development. It is a crucial ingredient in thyroid hormone, so a lack of iodine in the diet will reduce the thyroid's ability to manufacture thyroid hormone resulting into hypothyroidism. Symptoms of hypothyroidism include weight gain, inability to lose weight, fatigue, elevated blood lipids, hair loss, dry skin, loss of libido, infertility, to name a few. This results in the disease states collectively known as Iodine Deficiency Disorders (IDD). Cretinism; iodine deficiency during pregnancy often results in abnormal neurodevelopment and lower intelligence quotient (IQ) in the child.

Fluoride

Fluoride exists in soil as fluorapatite and fluorosilicate. It gets added in significant quantities to the soil through irrigation water and soil amendments and fertilizers like gypsum, phosphogypsum, apatite, phosphatic fertilizers. Fluoride is easily absorbed by plants and its deficiency has not been reported in plants.

It is essential for the normal growth and development of bones in human and animals. Fluorine has been recognized as an essential trace element because of its prophylactic effects in dental caries. Humans get most of their F from water that in turn comes from the soil as well as the underlying rocks.

Excessive quantities of F become poison both for humans and animals. Toxic concentrations of F interfere with Ca metabolism causing simultaneous osteosclerosis of the spine and osteoporosis of the limb bones. Harmful effects of excessively high natural concentrations of F in water and plants on human and animal health has been reported from many countries including India (Krishamuchari, 1973). Fluorine toxicity in human beings causes dental mottling and skeletal fluorosis. Fluoride toxicity, as dental fluorosis, expressed in the form of pigmentation with yellow, brown or black coloration, mottling, irregular wearing, erosion, pitting of enamel on teeth has been reported in sheep found around National Aluminum Company (NALCO) smelter plant in Orissa.

Chromium

Chromium (Cr) content in Indian soils is very low. Plants have ability to absorb chromium though it is not an essential nutrient. Its uptake is very small from the normal and unpolluted soils. Tetravalent chromium is toxic whereas hexavalent is essential and beneficial to human health.

In animals, adequate level of Cr has been found to increase growth and longevity.

Chromium is essential trace element for humans and plays an important role in human and animals mainly in regulation of the glucose tolerance factor, in combination with nicotinic acid and some proteins which are required for every bodily function. Deficiencies are believed to be a factor in arteriosclerosis and hypertension and possibly in diabetes and cataract. Lack of chromium is known to cause serious eye abnormalities (Shukla and Behera, 2018)

5.4 STRATEGIES TO IMPROVE MICRONUTRIENT NUTRITION IN ANIMALS AND HUMANS

Problems pertaining to deficiency and toxicity of micronutrients in animal and human health emanate from low and very high flow of these elements to animals and humans through the soil-water plant- animal-human food chain in different geographical areas. Soil is the major source supplying micro-nutrient elements to food chain. Cereals namely, rice and wheat grown in Zn- and Fe deficient soils produce grains in low Zn and Fe content. Consequently, the cereal-based food/diet, that contributes 70 % of the daily calorie intake of the poor population, is also low in Zn and Fe concentration. Following approaches are used to manage and/or prevent micronutrient deficiencies and improve their status in humans.

Diversifying Diets

Increasing dietary diversity means increasing both the quantity and the range of micronutrient-rich foods consumed. In practice, this requires the implementation of programmes that improve the availability and consumption of, and access to, different types of micronutrient-rich foods (such as animal products, fruits and vegetables) in adequate quantities, especially among those who at risk for, or vulnerable to, MNM. Increasing dietary diversity is the preferred way of improving the nutrition of a population because it has the potential to improve the intake of many food constituents – not just micronutrients – simultaneously.

Increasing dietary diversity is one of the most effective ways to sustainably prevent hidden hunger (Thompson and Amoroso 2011). Dietary diversity is associated with better child nutritional outcomes, even when controlling for socioeconomic factors. In the long term, dietary diversification ensures a healthy diet that contains a balanced and adequate combination of macronutrients (carbohydrates, fats, and protein); essential micronutrients; and other food-based substances such as dietary fiber. A variety of cereals, legumes, fruits, vegetables, and animal-source foods provides adequate nutrition for most people, although certain populations, such as pregnant women, may need supplements (FAO 2013). Effective ways to promote dietary diversity involve food-based strategies, such as home gardening and educating people on better infant and young child feeding practices, food preparation, and storage/preservation methods to prevent nutrient loss.

Fortifying Commercial Foods

Commercial food fortification, which adds trace amounts of micronutrients to staple foods or condiments during processing, helps consumers get the recommended levels of micronutrients. In many situations, this strategy can lead to relatively rapid improvements in the micronutrient status of a population, and at a very reasonable cost, especially if advantage can be taken of existing technology and local distribution networks. A scalable, sustainable, and cost-effective public health strategy, fortification has been particularly successful for iodized salt: 71 percent of the world's population has access to iodized salt and the number of iodine-deficient countries has decreased from 54 to 32 since 2003 (Andersson, et. al., 2012). Other common examples of fortification include adding B vitamins, iron, and/or zinc to wheat flour and adding vitamin A to cooking oil and sugar.

Fortification of food with micronutrients is a valid technology for reducing micronutrient malnutrition as part of a food-based approach when and where existing food supplies and limited access fail to provide adequate levels of the respective nutrients in the diet. In such cases, food fortification reinforces and supports ongoing nutrition improvement programmes and should be regarded as part of a broader, integrated approach to prevent MNM, thereby complementing other approaches to improve micronutrient status.

Supplementation

Supplementation is the term used to describe the provision of relatively large doses of micronutrients, usually in the form of pills, capsules or syrups. It has the advantage of being capable of supplying an optimal amount of a specific nutrient or nutrients, in a highly absorbable form, and is often the fastest way to control deficiency in individuals or population groups that have been identified as being deficient.

In developing countries, supplementation programmes have been widely used to provide iron and folic acid to pregnant women, and vitamin A to infants, children under 5 years of age and postpartum women.

Vitamin A supplementation is one of the most cost-effective interventions for improving child survival (Edejer et. al. 2005). Programs to supplement vitamin A are often integrated into national health policies because they are associated with a reduced risk of all-cause mortality and a reduced incidence of diarrhoea (Babar et. al. 2010). Because a single high-dose vitamin A supplement improves vitamin A stores for about 4–6 months, supplementation two or three times a year is usually adequate. However, in the case of the more water-soluble vitamins and minerals, supplements need to be consumed more frequently

Supplementation for other micronutrient deficiencies is less common. In some countries, iron-folate supplements are prescribed to pregnant women though coverage rates are often low and compliance rates even lower. For children, home fortification with micronutrient powders and lipid-based nutrient supplements can include multiple micronutrients, like iron and zinc, but they are harder to get into homes on a large scale than vitamin A supplements.

The control of vitamin and mineral deficiencies is an essential part of the overall effort to fight hunger and malnutrition. Eliminating hidden hunger will not be easy. Challenges lie ahead. But if enough resources are allocated, the right policies developed, and the right investments made, these challenges can be overcome. Countries need to adopt and support a comprehensive approach that addresses the causes of malnutrition and the often associated "hidden hunger" which rest intrinsic to in poverty and unsustainable livelihoods. Actions that promote an increase in the supply, access, consumption and utilization of an adequate quantity, quality and variety of foods for all populations groups should be supported. The aim is for all people to be able to obtain from their diet all the energy, macro- and micronutrients they need to enjoy a healthy and productive life.

5.5 BIOFORTIFICATION- PRODUCING MICRONUTRIENT-RICH FOOD

Over the last 50 years, agricultural research for developing countries has increased production and availability of calorically dense staple crops, but the production of micronutrient-rich non-staples, such as vegetables, pulses and animal products, has not increased in equal measure. Non-staple food prices have increased steadily and substantially, making it more and difficult for the poor to afford dietary quality. In the longterm, increasing the production of micronutrient-rich foods and improving dietary diversity will substantially reduce micronutrient deficiencies. In the near term, consuming biofortified crops can help address micronutrient deficiencies by increasing the daily adequacy of micronutrient intakes among individuals throughout the lifecycle (Bouis and Saltzman, 2017). Production of micronutrient-rich food involving genetic modification in crops and change in fertilizer strategies through innovative agricultural interventions such as biofortification, offers a sustainable solution to tackle micronutrient malnutrition in the poverty-ridden population.

Biofortification is a process of increasing the density of vitamins and minerals in a crop through plant breeding, transgenic techniques, or agronomic practices. Biofortified staple crops, when consumed regularly, will generate measureable improvements in human health and nutrition (Bouis and Saltzman, 2017).

Biofortification through the Mode of Minerals Fertilization

Biofortification entails the improvement in the nutritional quality of a target crop by enhancing the micronutrient concentration in edible portions without sacrificing agronomic traits, i.e., pest resistance, yield, drought resistance (Klikocka and Marks, 2018). Plant breeders also improve yield and pest resistance, as well as consumption traits, like taste and cooking time—to match or outperform conventional varieties.

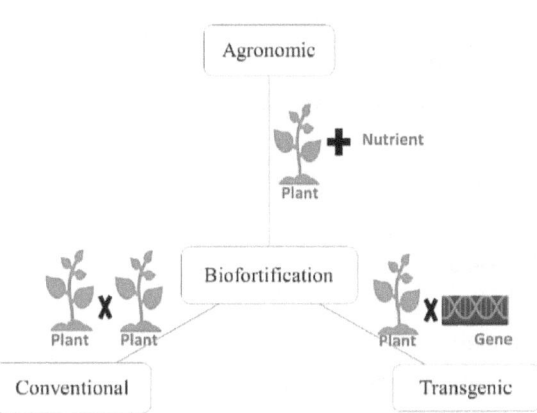

Fig. 5.3: Different approaches to achieve biofortification

Three major strategies to accomplish nutritional security through biofortification are conventional biofortification, transgenic biofortification and agronomic biofortification (Fig5.3). The conventional approach refers to selecting existing varieties of high-yielding crops that naturally contain a higher content of nutrients of interest and cross-breed using conventional methods to produce staple crops with desirable nutrient and agronomic traits (Dhaliwal *et. al.*, 2022).

Genetic Biofortification

Genetic biofortification can be attained through specific genetic manipulation to enrich micronutrient concentration in edible plant parts. is a seed based approach where the germ plasm is enriched with specific nutrients: micronutrients, protein, amino acids etc. It can be done by conventional breeding, marker driven molecular breeding, or genetic engineering. Some of the promising products of this approach are Zn- and Fe- rich rice, wheat and maize. Golden rice rich in beta-carotene, high Fe rice (high ferritin gene from mangroves) are examples of genetic engineering for bio fortification (Shukla *et. al.* 2018).

Some of the promising products of this approach are Zn- and Fe- rich rice, wheat and maize. Golden rice rich in beta-carotene, high Fe rice (high ferritin gene from mangroves) are examples of genetic engineering for bio fortification (Shukla *et. al.* 2018).

Agronomic Biofortification

Non-genetic measures or agronomic biofortification to enhance the micronutrient level in food plants could be more efficient. Indeed, this approach not only improves yields but also the nutritional quality of grains (Cakmak and Kutman, 2017). Agronomic biofortification refers to the micronutrient fertilizers applied to the soil

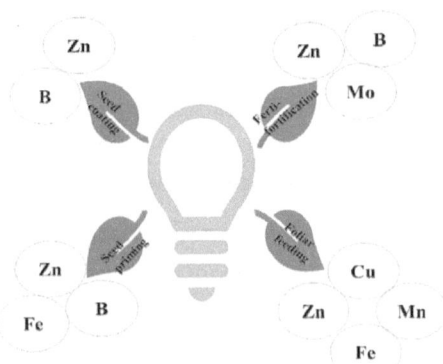

Fig. 5.4: Different modes of mineral fertilization.

and/or plant to enrich the edible part of the field crop with micronutrients. Agronomic biofortification is an inexpensive and simple approach which can be utilized to enrich the genetically-inefficient cultivars by application of micronutrient fertilizers at different rates, methods and at different crop growth stages (Shukla *et. al.* 2018).

Of the several strategies developed using permutation and combinations of nutrient management options, soil application with foliar feeding is the best for grain enrichment. Soil application ensures a sufficient level of nutrient for root uptake, whereas foliar application increases the nutrient content in the leaf for its transportation to other plant parts. Kumar *et. al.*, 2018 reported the agronomic intervention as a sustainable and cost-effective approach to improve plant growth, yield contributing traits and Zn, Fe content in paddy crops using N, Zn and Fe fertilization. Biofortified crops that have been released so far include vitamin A orange sweet potato, vitamin A maize, vitamin A cassava, iron beans, iron pearl millet, zinc rice, and zinc wheat. While biofortified crops are not available in all developing countries, biofortification is expected to grow significantly in the next five years.

DIFFERENT MODES OF MINERAL FERTILISATION

Ferti-Fortification or Soil-Applied Fertilizers

Owing to multiple benefits, micronutrients application through soil is the most versatile and effective method, particularly to correct the deficiencies of B (Dhaliwal and Manchanda, 2009; Dhaliwal, *et. al.*, 2012). To alleviate Zn deficiency in different crops, soil application of $ZnSO_4.7H_2O$ (21% Zn) at 62.5 kg/ ha or $ZnSO_4H_2O$ (33% Zn) at 40 kg /ha have been found equally efficient and economical.

The data on soil application of minerals has been represented in Table 5.4.

Table 5.4 Ferti-fortification of minerals in various crops

Micronutrient	Crop	Reference
Zinc	Wheat	Dhaliwal, *et. al.*, 2012
Zinc	Rice, Wheat	Ram *et. al.*, 2015
Zinc	Cowpea	Manzeke *et. al.*, 2017
Zinc	Lentil	Rasheed *et. al.*, 2020
Zinc + Selenium	Soybean	Dai *et. al.*, 2020

Foliar Feeding

Soil fertilization can enhance the micronutrient levels within the grain; however, it limits the salt intake and is not efficient for immobile minerals. Moreover, the micronutrients such as Fe present in the fertilizer become futile for soil application, as soluble Fe is readily converted to insoluble Fe3+ form, thus making it unavailable to plants. Thus, foliar feeding of fertilizers was implemented to overcome such problems. This intervention involves the fertilizers spray in the liquid state applied over the leaves. A single or mixture of micronutrient solutions in combination with the target salt is applied as a spray on the leaves where they are absorbed via stomata and epidermis. Thus, in particular cases, foliar feeding evidenced more efficient as compared to soil applications for effective use of nutrients and minimizing the visual deficiency and soil deficiency problems shortly. To date, a wide range of studies has been reported on foliar feeding of mineral fertilizers to different crops (Table 5.5)

Table 5.5. Foliar application of micronutrients for mineral fertilization.

Micronutrient	Crop	Reference
Zn and Fe	Wheat	Dhaliwal et. al., 2009
Zn and Fe	Rice and Brown rice	Dhaliwal et. al., 2010
Mn and Cu	Wheat	Dhaliwal et. al., 2011
Fe	Rice	Singh et. al., 2013
Fe and Zn	Maize	Dhaliwal et. al., 2013
Zn	Wheat and Rice	Ram et. al., 2015
Zn + Mn	Wheat, rice, barley	Amanullah et. al;., 2018
Cu	Oats	Sandhu et. al., 2020
Fe	Rice	Kumar et. al., 2016
Zn	Oats	Dhaliwal et. al., 2020
Zn	Chickpea	Singh et. al., 2019

Seed Priming

Seed priming involves controlled hydration of seeds that permits them to perform their pre-germination metabolic events without radical emergence. Farooq *et. al.*, 2006 reported that primed seeds have immense potential to give uniform stand establishment and productivity than dry seeds (Table 5.6).

Seed Coating

Seed coating involves the application of nutrients in the form of finely ground powder to the seed surface along with the inert sticky material (e.g., Arabic gum). This technique affects the seed or soil at the soil–seed interface and ultimately influences the availability of soil-applied and coated nutrients. The benefits of seed coating include uniform application of micronutrient source and reduction in the adverse effect on non-target pests, thus minimizing the environmental side effects. Moreover, the combination of treatments can be applied more precisely. Numerous factors including the nutrient: seed ratio, coated micronutrient, soil type, soil fertility, soil moisture and material used for coating alter the proficiency of micronutrients.

Table 5.6 Seed treatment with various micronutrients

Seed Treatment	Micronutrient	Crop	Reference
Seed Priming	B	Wheat and Rice	Atique et. al., 2012
	Zn + Pseudomonas sp. MN12	Bread-wheat	Rehman et. al., 2018
	Fe and Zn	Wheat and Barley	Caravalho et. al., 2019
	Zn	Wheat	Rehman and Farooq, 2016
Seed Coating	B	Rice	Rehman and Farooq, 2013
	Mn	Bread-wheat	Amanullah et. al;., 2018

BIOFORTIFICATION THROUGH AGRONOMIC MANAGEMENT PRACTICES

Management practices are primarily focused on improving the soil condition that ultimately enhances the micronutrient availability for plant uptake and thus its concentration in food. Various factors such as physical, chemical and biological characteristics of soil determine the nutrient use efficacy of the plant. Well-known management practices are integrated soil fertility management (IFSM), tillage practices and water management. The integrated soil fertility management approach includes the use of mineral fertilizer, organic inputs and improved germplasm (Selim, 2020). Apart from the soil organic matter sustainability, organic compounds advocate multiple benefits to soil status, such as water holding capacity, cation exchange capacity and soil structure. However, they release nutrients in the rate determined manner, thus seldom meeting the crop nutrient requirement at the adequate time for optimum yield. Thus, sole usage of organic inputs and mineral fertilizers are not sufficient to cover the gap between mineral deficiency and the requirement of grain crops. Thus, the combination of both interventions is effective as they have positive interactions and complementary functions (Padbhushan, *et. al.*, 2021). With the increase in agricultural activities, various approaches for soil amendments have been frequently used and proved beneficial.

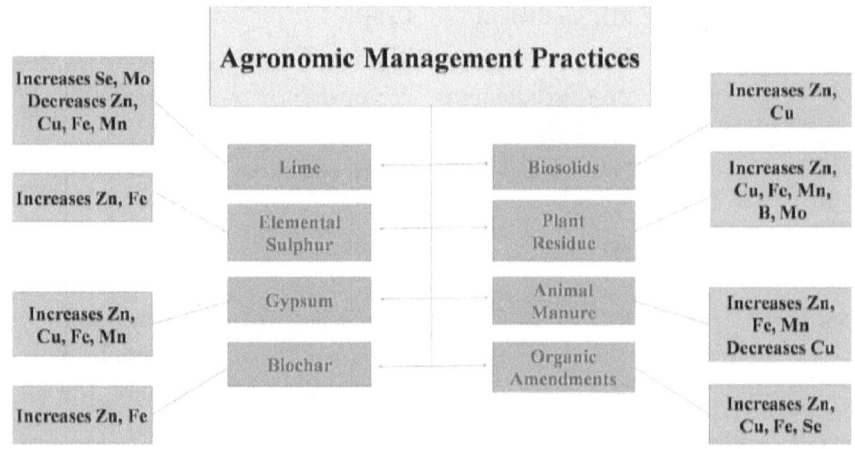

Fig. 5.5: *Different amendments under agronomic management practices to enhance mineral content in plant*

Among the chemical soil properties, the soil pH plays a pivotal role in affecting plant growth, improving nutrient availability and eradicating harmful toxic substances. Thus, optimizing soil pH is a prerequisite to accomplishing quantitative and qualitative crop production. It involves either increasing the acid soil pH or lowering the pH of alkaline/calcareous soils.

No single intervention will alleviate micronutrient deficiencies, and biofortification complements existing interventions, such as supplementation and industrial food fortification. Biofortification, however, has two key comparative advantages: its long-term cost-effectiveness and its ability to reach underserved, rural populations. Once developed, nutritionally improved crops can be evaluated and adapted to new environments and geographies, multiplying the benefits of the initial investment. Once the micronutrient trait has been mainstreamed into the core breeding objectives of national and international crop development programs, recurrent expenditures by agriculture research institutes for monitoring and maintenance are minimal.

Biofortified crops are also a feasible means of reaching rural populations who may have limited access to diverse diets or other micronutrient interventions. Target micronutrient levels for biofortified crops are set to meet the specific dietary needs of women and children, based on existing consumption patterns. Biofortification puts a solution in the hands of farmers, combining the micronutrient trait with other agronomic and consumption traits that farmers prefer. After fulfilling the household's food needs, surplus biofortified crops make their way into rural and urban retail outlets (Bouis and Saltzman, 2017).

5.6 FIELD CROPS AS A RICH SOURCE OF MINERALS

To eradicate malnutrition, the dietary pattern must incorporate the minimum required amount of essential nutrients including carbohydrates, fat, proteins, vitamins and minerals (Lean, 2019). To meet this requirement, various types of crops are grown globally, including cereals, pulses, oilseeds and fodder crops. Due to the lack of a diverse diet in developing countries, the staple crops must contain essential nutrients in an adequate diet. Among them, rice, maize (corn) and wheat contribute to two-thirds of world staple crops intake. Apart from them, millet and sorghum, cassava, potatoes, yams, taro, dairy products and animal products such as meat, fish are major dietary intakes. The principal diet in low-income developing countries constitutes cereals as staple food, including wheat, while rice, millet, corn and sorghum.

CEREALS

Apart from global food dependency, cereal grains can absorb more mineral nutrients in their edible parts. Thus, the scientific community emphasizes improving the profile of cereal grains concerning micronutrients. Among cereal grains, rice is consumed by more than half of the world population and contributes to more than 42% of calorie intake (Huang *et. al.*, 2020). About 90% of global rice consumption is in Asia and is the fastest-growing staple food in Africa and Latin America. The nutritional composition of rice consists of several vitamins (such as vitamin B1, B2, B3, B6), diverse phyto-molecules and various minerals such as Na, Zn, Fe, Cu, K, Mg and P.

The second staple food after the rice is wheat, in South Asia, Turkey and China, and contributes to more than 70% of daily calorie intake. Thus, its nutritional value is of utmost importance to meet nutritional security. Its diverse functionality enhances its importance as its seeds can be grounded into semolina, flour, etc., which are the basic ingredients of pasta, bread and bakery products.

Corn, also termed as maize, is the world's leading staple crop along with wheat and rice. Its popularity is largely due to its use as a food source for both humans and animals. It can be consumed in many ways, such as boiled, roasted, fried, ground and fermented, for use in gruel, bread, cakes, porridges and alcoholic beverages. In industries, its use is in sweeteners, oils, food thickeners and non-consumables. Its nutritional content consists of 72% starch, 10% protein and 4% lipid. Maize contains numerous important vitamins except for vitamin B-12. The most prevalent minerals found in maize are K, Mg, Zn and P (in the form of phytate). Sulfur is present in the form of methionine and cystine. Ca and Fe are found in negligible amounts and other trace minerals include Mn, Se, Cu and I (Nuss and Tanumihardjo, 2010).

PULSES

Pulses are also termed as a superfood due to their intense nutritional composition providing multiple health benefits. Their acknowledgment to achieve nutritional security can be recognized as they were declared the theme of the year 2016 by FAO (Food and Agricultural Organization of the United Nations). After cereals, they rank as the second major crop production in the agricultural sector worldwide, with major consumption (over 70%) of lentils, peas, beans and chickpeas worldwide (Calles *et. al.*, 2019).

Globally, six pulses, namely, chickpea (*Cicer arietinum*), field pea (*Pisum sativum*), lupin (*Lupinus*), faba/broad bean (*Vicia fabae*), lentil (*Lens culinaris*) and mung bean (*Vigna radiata*), have gained importance due to their major production (Calles *et. al.*, 2019) India is the leading producer worldwide, followed by Canada, Myanmar and China. Notably, they are enriched with protein up to 30% of their weight on a dry basis), carbohydrates, dietary fibre, energy, important bioactive compounds and essential minerals and vitamins required for human health (Rebello *et. al.*, 2014). Among the pulses, lentils and beans have the highest Fe and Zn contents. The Fe and other minerals content are generally highest in beans. However, the presence of phytate and other inhibitors lower the digestibility or bioavailability of nutrients; thus, its removal is necessary. After its degradation, pulses would become potential sources of Zn and Fe due to immense mineral content (Venkidasamy *et. al.*, 2019). The folate concentration of beans accounts for 95% of daily requirements.

OILSEEDS

Oilseed crops are grown primarily for the extraction of oil contained in the seeds. Vegetable oils are used worldwide for household cooking and in other food products such as baked items and snacks. Apart from this, they are used as raw materials for various oleo-chemical industries. Oilseeds include canola, safflower, corn, sunflower, olive, soybean and peanut crops. As compared to other crops, oilseeds are considered a major vegetable oil source with a higher yield of oil (Zafar *et. al.*, 2019). In the semi-arid tropical regions, these crops are particularly important among low-income families as they contribute 40% of the total calorie intake in their diet. Recent reports confirmed the high protein and mineral-like Zn, Fe, Ca, etc. content in rape oilseed, thus making them the potential alternative to traditional crops for tackling nutritional stress (Kowalska *et. al.*, 2020).

Soy hulls are rich sources of minerals, fibre, protein and energy, with low-lignin content, and are highly palatable. Canola meal contains proteins, amino acids, vitamins and minerals with high sulfur content. Hemp seed and mustard seed oil are enriched with an adequate supply of nutritional parameters and several minerals including Ca, Mg, P, S, K, along with Fe and Zn. Likewise, other oilseeds are highly beneficial for human nutrition.

The major constraint in oilseeds is the high level of phytic acid and other binding agents which reduce mineral bioavailability from the seeds. Therefore, major emphasis is given to developing the methodology for phytate removal to control or reduce the mineral binding in oilseed products. On the other hand, oilseed enriched with minerals via biofortification intervention may alleviate the mineral deficiency. The improved Se content in the plant components of oilseed rape has been observed on foliar application of Na_2SeO_4 under acidic location (Száková *et. al.*, 2017).

FODDER

Humans depend on livestock directly or indirectly for their nutrition. Thus, nutrient deficiency in animals or their food will eventually affect the food of the end-user, i.e., human. Like humans, animals too need an adequate diet for optimal growth and productivity. However, plant food alone is insufficient in order to cover the gap of mineral intake; thus, the addition of mineral supplementation is required in animal feeds (Lopez-Alonso, 2012).

Examples of forage crops include cowpea, sorghum, maize, Guar, Soybean, Berseem, Lucerne etc. In animals, two essential elements can be characterized as macro and microelements based on their daily intake required. Macro-elements include Ca, P and K, while Mn, Fe, Cu, Zn and Mo are part of the microelements category. Some non-essential plant elements are those such as Na, Se, Cr, I, Si and V, but essential elements for humans and animals make their way in plants via non-selective transport mechanisms, thus entering into humans and livestock food supply.

Recently, biofortification of fodder crops has gathered researcher attention due to their economic importance and agricultural trends globally (Jadhav et. al., 2020). A wide range of new cultivars involving *Medicago* spp., *Trifolium* spp., *Lolium*, and *Festuca* have been engineered to improve their biomass production, nutritional values, digestibility of DM, resistance against biotic stress (fungal and nematodes) and crop durability (Boller and Greene, 2010). efficient biofortification has been achieved on temperate cultivars of Lolium and some tropical grasses including *Pennisetum* and *Brachiaria* (Cabral et. al., 2015).

Soil-applied Cu @ 6 kg/ha to oats fodder enhanced yield and Cu content due to increased bioavailability of Cu (Kaur et. al., 2015). Sandhu et. al.2020 studied the biofortification of oats fodder with foliar-applied Cu. Foliar-applied Cu (0.2%) at 60 and 90 days of sowing significantly enhanced the Cu uptake, content, yield and protein content.

Many genes have been identified which can be used to enhance forage crops. However, due to limited transformation technology, the development of new forage crops is lagging. Genome editing is a promising technology and will be more accepted in the future for forage crop development.

Impact of fertilizers with special reference to micronutrients on environment pertaining to human and animal health

Fertilizers are an indispensable agri-input and are added to aid and increase the supply of essential nutrients required for growth and development of plants. There are 16 nutrient elements required to growcrops. Three essential nutrients - carbon (C), hydrogen (H), and oxygen (O) - are taken up from atmospheric carbon dioxide and water. The other 13 nutrients are taken up from the soil and are usually grouped as primary nutrients, secondary nutrients, and micronutrients. The primary nutrients are nitrogen (N), phosphorus (P), and potassium (K). Primary nutrients are utilized in the largest amounts by crops, and therefore, are applied at higher rates than secondary nutrients and micronutrients. The secondary nutrients - calcium (Ca), magnesium (Mg), and sulphur (S) - are required in smaller amounts than the primary nutrients, although there are opinions that S may be a major nutrient in some cases.

Micronutrients - iron (Fe), manganese (Mn), zinc (Zn), copper (Cu), boron (B), and molybdenum (Mo), selenium (Se) -are required in even smaller amounts than secondary nutrients. They are available as manganese, zinc and copper sulphates, oxides, oxy-sulphates

And chelates, as well as in boric acid and ammonium molybdate. Entire focus on use of fertilizers have been almost exclusively on NPK since their "discovery" in the mid-1800s. Owing to the role of these primary nutrients in the crop cycle, they have been overused to increase the production without realizing the long-term effects on the soil and environment. Thus, the negative impact of fertilizers has often been highlighted. But as the fertilizers are required for proper growth of plants, and plants are the primary source of food for human and animal consumption, the former does have an impact on human and animal health as well.

Impact of micronutrients on human health

As seen in the case of micronutrients, mineral elements like Zn, Fe and Cu are as important as compounds like carbohydrates, fats, protein and vitamins for human development. Micronutrient deficiencies in soils limit crop yields and nutritional quality, which in turn negatively affect human health (Marschner, 2012; Alloway, 2009) Micronutrient intake less than the recommended values can cause slower physiological processes. High consumption of cereal-based foods with low contents of micronutrients is causing health hazards in humans (Imtiaz., 2010).

Micronutrient malnutrition, the so-called hidden hunger, affects more than one-half of the world's population, especially women and preschool children in developing countries (Welch and Graham, 2004). Worldwide over 2 billion people suffer from iron (Fe), zinc (Zn) and/or other (multiple) micronutrient deficiencies (WHO, 2016; Black, 2003). The problem is most severe in low and middle income countries (Muthayya., 2013). The physiological impacts of micronutrients are complex, relating to many bodily functions. Even mild to moderate deficiencies of micronutrients can lead to severe health problems.

Selenium: Selenium has important antioxidant, anti-cancer and antiviral properties and its deficiency makes human prone to thyroid dysfunction, cancer, severe viral disease and various inflammatory conditions (Lyons et. al., 2004).

Zinc and Iron: Iron deficiency is the prominent cause of anaemia which contributes to compromised physical productivity, cognitive impairment and adverse pregnancy outcomes. Likewise, Zn deficiency has been related to growth failure, decreased immunity leading to increased susceptibility to infection, morbidity and mortality due to diarrheal disease, and the incidence of respiratory tract pneumonia (Etcheverry et. al., 2005), impaired growth and development of infants, children and adolescents, as well as impaired maternal health and pregnancy outcome (Martin, 2004).

Trials conducted in several countries indicate that duration and severity of major baby-killers such as diarrhoea and pneumonia can be reduced by 30-50 % by supplying adequate amounts of vitamin A and zinc (Bhargava et. al., 2001). In developing countries, zinc deficiency ranks 5th among the leading 10 risk factors. Even on a global scale, taking developed and developing countries together, zinc deficiency ranks 11th out of the 20 leading risk factors. WHO attributes 800,000 deaths worldwide each year to zinc deficiency and over 28 million healthy life years lost. It is estimated that zinc deficiency affects one-third of the world's population, with estimates ranging from 4 to 73% according to regions, and it is 5th leading risk factor along with the Fe deficiency, the latter is at 6th position globally.

Advanced technology to mitigate harmful effect of micronutrients and on its content in foodstocks

Chelation as an efficient delivery system

Chelation is the most advanced delivery system for crop nutrition which converts metal nutrients into inert, water soluble compounds, thus making the crop nutrients incapable of any harmful chemical reactions once applied to the soil and crops. This protects the fields from harm caused by inorganic fertilizers.

The challenge of integrated nutrient management may be addressed by creating a range of specialty plant nutrient products that are chemically inert in nature. This process is known as Chelation Technology and has been successfully adapted to the agricultural sector. The process involves taking positively charged metal nutrients like zinc, iron, copper, manganese, magnesium, etc. and enveloping these ions with a negatively charged chelating agent. Once this process is complete, the end-product is a stable complex, which is chemically inert until it enters the plant system. The enzymes inside the plant system reactivate the product and release the nutrients to the crop. Thus, chelation is a delivery system which leads to the nutrients being delivered to the plant in its most unharmed form.

Post-delivery of the nutrients, the chelation agent is released back into the root zone and it attracts heavy metals (positively charged metals) like lead, arsenic, cadmium, etc. and neutralizes these toxins in the soil thereby rendering them inactive and incapable of causing further harm to the crop. This residual effect is an added advantage of chelation technology.

Specific formulations containing various nutrients required by different crops based on soil, crop and geographic conditions have also been created. This makes it very simple for the farmer to access specific formulations as per crop requirement. Chelation in India largely uses synthetic chelating agents like EDTA, DTPA, EDDHA, etc. However, soya protein also provides a safe, cost-effective natural alternative to providing chelated nutrients to crops.

Economic benefits of chelation technology

Chelates remain inert till they are absorbed into the plant system. There is zero wastage and a reduction in dosage upto 20 times compared with the traditional inorganic nutrient application. Thus, a farmer applies only 500 g of a chelate versus 20 kg of a sulphate of the similar nutrient. Despite the low dose of the chelate, yields are reported to increase by an average of 30%. Cost-benefit consequently is very favourable, with a farmer gaining on average Rs 6/- for every Re1/- invested in chelated nutrients, experimentally observed.

Improving micronutrient content in foodstocks

Micronutrient content of food can be increased either by supplementation, fortification or by agricultural management strategies. The process of food fortification and supplementation are too expensive and hence impractical to be applied on large scale, hence not easily accessible to poor masses. A suggested strategy to alleviate micronutrient deficiencies is agronomic biofortification, particularly of staple foods, and application of micronutrients containing fertilizers.

Genetic biofortification involves either genetic engineering or classical breeding. Agronomic biofortification is achieved through micronutrient fertilizer application to the soil or application directly to the leaves of the crop (foliar application). The impact of agronomic biofortification largely depends on the bioavailability of micronutrients throughout the pathway from soil to plant, from plant to food, and uptake by the human body (Fig1)

Interactions between micronutrients and macronutrients can influence the effectiveness of agronomic biofortification. Adequate N and P status of plants has a positive influence on root development, shoot transport, and re-localization of nutrients from vegetative tissue to the seeds (Prasad et. al., 2014; Cakmak et. al., 2010). This results in increased micronutrient uptake and concentrations in the edible parts of the crop. There are several critical factors that play a key role to determine the success of agronomic fortification to overcome micronutrient deficiencies among humans. These factors depend on nutrient bioavailability at different stages: the presence and bioavailability of soil nutrients for plant uptake, nutrient allocation within the plant, re-translocation into the harvested food, and availability of nutrients in prepared food for uptake in the human body. Bioavailability from soil to crop is influenced by many soil factors (viz, pH, organic matter content, soil aeration and moisture and interaction with other elements) and the crop variety that, for example, defines the functioning of rooting systems. Bioavailability from crop to food is influenced by the crop variety (which defines the allocation and re-localization of micronutrients into edible parts of the crop) and processing of the harvestable part (i.e., milling and dehusking). Bioavailability from the food for the human body is influenced by cooking of food and dietary intake such as, the amount of food consumed, diet composition and individual health status (deValença and Bake, 2016).

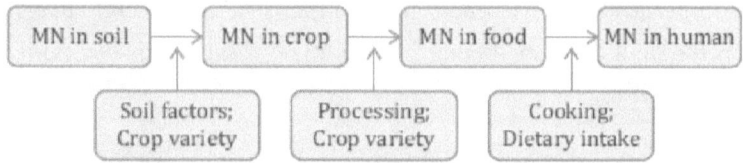

Fig. 1. Schematic overview of micronutrient pathway from soil to humans and the factors that influence its bioavailability (Mayer et. al., 2011)

Impact of micronutrients on animal health

In animals, micronutrients are required in the diet for their overall improved health and also essential for the production of egg, meat and milk. The importance of micronutrients can be very well realized by the fact that their deficiencies in animal diet can lead to restricted growth and reduction in animal productivity. Fe, Zn, Cu, Mn, Mo, Se, I and Co are the essential elements which play pivotal role in animal health and each element play at least a major role in physiological functioning of animal (Table 1).

Table 1. Deficiency symptoms and the major role by micronutrients in ruminant livestock (Fischer, 2008)

Element	Role	Deficiency symptom / Diseases
Fe	Protein and enzyme function. Blood haemoglobin.	Anemia
Cu	Haemoglobin formation, enzyme function, and pigments	Anaemia, poor growth, bone disorders, infertility, brain and spinal cord lesions. Decolouration of hair.
Co	Vitamin B12 function and energy assimilation.	Poor growth, anaemia, loss of coat, low immunity to disease, infertility
Se	Vitamin E function	Poor growth, white muscle disease, infertility
I	Thyroid gland function	Goitre and reproductive failure
Mn	Enzyme activation	Enzyme activation
Zn	Enzyme function	Stiff and swollen joints, parakeratosis.
B	Enzyme function	Weak bones, poor immune function

It is imperative to have a balanced and holistic approach to overcome micronutrient deficiencies in practical farm situations as mentioned by Fischer (2008).

- *Treating the soil with fertilizers and nutrients particularly in which the soil and animal is likely to be deficient.*
- *Treating the fodder or herbage with micronutrients through foliar spray.*
- *Treating the animals by using feeding blocks and licks*
- *Supplementation of micronutrients through feed.*
- *Directly injecting the animals with nutrients*

Effects and Inferences

When micronutrient demand and supply are in sync, there should be no serious negative environmental effects seen within the agricultural ecosystem. Micronutrients generally bind strongly to the soil and thus are not susceptible to loss into the environment which minimizes risks of environmental pollution. Furthermore, micronutrients improve crop health, which reduces the need for agrochemicals (pesticides, herbicides, fungicides, etc.). Accumulation in soils due to overuse may cause toxicity problems. The globally available mineral reserves of micronutrients are limited, which highlights the importance of nutrient recycling for long-term sustainable micronutrient availability for agricultural production.

Rahul Mirchandani and Shama Zaidi
Fertilizers and Environment News, Vol.4, No. 2

Recent Advances In Agriculure Technology

6.1 RECENT ADVANCES IN AGRICULTURE

Much has changed in the field of agriculture over the years. While some traditional methods and technologies remain, more efficient, effective, and innovative technological advancements in agriculture are creating new opportunities and transforming how farming is done. This has changed how crops are grown and led to more efficient methods of resource management. Innovation is more important in modern agriculture than ever before. The industry as a whole is facing huge challenges, from rising costs of supplies, to shortage of labor, and changes in consumer preferences for transparency and sustainability. There is increasing recognition from agriculture corporations that solutions are needed for these challenges.

Today, the impact of technology on agriculture is undeniable. Engineers and researchers are continuously working hard to develop new technologies that solve farming, crops, and livestock management problems. Major technology innovations in the space have focused around areas such as indoor vertical farming, automation and robotics, livestock technology, modern greenhouse practices, precision agriculture and artificial intelligence, and blockchain.

Listed below are some technological advancements that are making a big impact in agriculture:

PRECISION AGRICULTURE

Agriculture is undergoing an evolution - technology is becoming an indispensable part of every commercial farm. New precision agriculture companies are developing technologies using GPS and other technological tools that allow farmers to maximize yields by controlling every variable of crop farming such as moisture levels, pest stress, soil conditions, and micro-climates. By providing more accurate techniques for planting and growing crops, precision agriculture enables farmers to increase efficiency and manage costs.

This is one of the most widely used technological advancements in agriculture, especially in large-scale farming, where every input matters. Farmers who embrace precision farming see higher yields, better soil health, and improved environmental impact. For instance, by using available tech to monitor soil health, farmers can avoid overfertilizing the land, which can be wasteful and cause disease.

FARM AUTOMATION AND ROBOTICS

Farm automation, often associated with "smart farming", is technology that makes farms more efficient and automates the crop or livestock production cycle. An increasing number of companies are working on robotics innovation to develop drones, autonomous tractors, robotic harvesters, automatic watering, and seeding robots. This technological advancement in agriculture has allowed farmers to increase yields of agricultural produce by increasing efficiency on farmlands. They can now use drones to map crops, monitor crop growth, and improve irrigation systems.

Drones are also used for aerial surveys to get a bird's eye view of the land, assess fallow fields or monitor irrigation levels across large areas. More farmers are turning to drones to map out their land for optimal grow times, crop rotation schedules, and harvesting needs. In livestock farming, robotics have also allowed the development of machines that can milk cows, shear sheep, and more.

Although these technologies are fairly new, the industry has seen an increasing number of traditional agriculture companies adopt farm automation into their processes. The primary goal of farm automation technology is to cover easier, mundane tasks. Farm automation technology addresses major issues like a rising global population, farm labor shortages, and changing consumer preferences. The benefits of automating traditional farming processes are monumental by tackling issues from consumer preferences, labor shortages, and the environmental footprint of farming.

LIVESTOCK FARMING TECHNOLOGY

The traditional livestock industry is a sector that is widely overlooked and under-serviced, although it is arguably the most vital. Livestock management has traditionally been known as running the business of poultry farms, dairy farms, cattle ranches, or other livestock-related agribusinesses. Recent trends have proven that technology is revolutionizing the world of livestock management. New developments in the past 8-10 years have made huge improvements to the industry that make tracking and managing livestock much easier and data-driven. This technology can come in the form of nutritional technologies, genetics, digital technology, and more.

Livestock technology can enhance or improve the productivity capacity, welfare, or management of animals and livestock. The concept of the 'connected cow' is a result of more and more dairy herds being fitted with sensors to monitor health and increase productivity. Putting individual wearable sensors on cattle can keep track of daily activity and health-related issues while providing data-driven insights for the entire herd.

Animal genomics can be defined as the study of looking at the entire gene landscape of a living animal and how they interact with each other to influence the animal's growth and development. Genomics help livestock producers understand the genetic risk of their herds and determine the future profitability of their livestock. By being strategic with animal selection and breeding decisions, cattle genomics allows producers to optimize profitability and yields of livestock herds.

Sensor and data technologies have huge benefits for the current livestock industry. It can improve the productivity and welfare of livestock by detecting sick animals and intelligently recognizing room for improvement. Data-driven decision making leads to better, more efficient, and timely decisions that will advance the productivity of livestock herds.

BLOCKCHAIN

Food traceability has been at the center of recent food safety discussions. Due to the nature of perishable food, the food industry at whole is extremely vulnerable to making mistakes that would ultimately affect human lives. When foodborne diseases threaten public health, the first step to root-cause analysis is to track down the source of contamination and there is no tolerance for uncertainty. Consequently, traceability is critical for the food supply chain. The current communication framework within the food ecosystem makes traceability a time-consuming task.

Blockchain in the supply chain has the potential to improve supply chain transparency and traceability as well as reduce administrative costs. The structure of blockchain ensures that each player along the food value chain would generate and securely share data points to create an accountable and traceable system. As a result, the record of a food item's journey, from farm to table, is available to monitor in real-time.

Blockchain's capability of tracking ownership records and tamper-resistance can be used to solve urgent issues such as food fraud, safety recalls, supply chain inefficiency and food traceability in the current food system. Blockchain's unique decentralized structure ensures verified products and practices to create a market for premium products with transparency. It also adds value to the current market by establishing a ledger in the network and balancing market pricing. The traditional price mechanism for buying and selling relies on judgments of the involved players, rather than the information provided by the entire value chain. Giving access to data would create a holistic picture of the supply and demand. Blockchain enables verified transactions to be securely shared with every player in the food supply chain, creating a marketplace with immense transparency.

GENETICALLY MODIFIED CROPS

Genetically modified crops are one of the most significant technological advancements in the agricultural sector. These types of plants have been altered to contain specific traits that will benefit farmers and consumers alike. They offer lots of benefits for farmers producing specialty crops like fruits and flowers. These include increased resistance to pests and diseases, tolerance to herbicides, better nutritional value, and resilience to adverse weather conditions.

GMOs have significantly reduced the amount of pesticides that farmers need to spray on their farms by up to 8.2% while increasing crop yield by 22%. This technological advancement in agriculture may not always be popular with consumers, but the science is clear – they're a safe and valuable tool for farmers. Planting GM crops also helps preserve soil, reduce carbon emissions, and conserve water.

ARTIFICIAL INTELLIGENCE

The rise of digital agriculture and its related technologies has opened a wealth of new data opportunities. Remote sensors, satellites, and UAVs can gather information 24 hours per day over an entire field. These can monitor plant health, soil condition, temperature, humidity, etc. The amount of data these sensors can generate is overwhelming, and the significance of the numbers is hidden in the avalanche of that data.

The idea is to allow farmers to gain a better understanding of the situation on the ground through advanced technology (such as remote sensing) that can tell them more about their situation than they can see with the naked eye. And not just more accurately but also more quickly than seeing it walking or driving through the fields.

REMOTE MONITORING OF CROPS USING SENSORS

Remote monitoring of crops using sensors such as drones and satellites is becoming increasingly popular. This allows farmers to monitor their fields from home, improving productivity by catching problems earlier and allowing for more efficient use of water and fertilizers. Crop sensors enable farmers to monitor their crops remotely from anywhere in the world using an app or web browser.

With such technological advancement in agriculture, farmers save on labor costs and increase their crop yields, making it possible to end food scarcity. Remote monitoring of crops using sensors is not only for large-scale farmers but also for smallholder farmers. A recent study showed that remote sensing could improve the accuracy of yield predictions by smallholder farmers in Africa by up to 30%. This will help these farmers make better decisions about their farming practices.

6.2 REAL-TIME ASSESSMENT OF NUTRIENT STATUS

6.2 a Field Sensing

In order to manage nutrients and to be able to take immediate action to correct deficiencies during the growing season, it is helpful if a quick and inexpensive means of determining nutrient deficiencies in the field is available. Visual observation through regular field scouting is perhaps the best approach. Knowing the normal visual characteristics (color, developmental stages, and morphology) of healthy plants and the ability to identify abnormalities is step one. It is very effective and inexpensive, and can be used in broad-acre high-tech production systems as well as small-plots systems managed manually. Various visual aids are available to help identify specific nutrient deficiencies for individual crops or for plants in general. The deficiencies often are caused by failure of a particular plant function that is affected and location reflects whether the nutrient is mobile in the plant (translocated from older to younger plant tissues) or immobile (not translocated).

The nutrient's role in the plant will determine if visual symptoms are possible as a diagnostic tool. The colors and patterns of the symptoms can be helpful in diagnosing these problems during the growing season. It is always helpful to confirm the diagnosis with plant analysis and soil tests.

THE LEAF COLOUR CHART

The LCC is an inexpensive and simple tool to monitor leaf greenness and guide the application of fertilizer N to maintain optimal leaf N content. Leaf color charts with four or six panels, ranging in color from yellowish green to dark green, have been developed. A big advantage of the LCC is that it is inexpensive and easy to understand and use. The LCC can be used to determine the relative rate of N fertilizer needed by the rice, wheat and maize crops; or it can be used to determine the timing for N fertilizer application.

The leaf N content is closely related to photosynthetic rate and biomass production, and can serve as an indicator of N status of the crop during the growing season. Leaf N content is reflected in the relative greenness of a leaf. Dark green leaves indicate enough N, whereas yellowish leaves indicate N deficiency. Therefore, the LCC is being successfully used to guide fertilizer applications, particularly in several Asian countries.

The LCC is used to manage fertilizer N starting from six-leaf (V6) stage up to R1 stage by applying a prescribed dose of N whenever leaf color was found to be less greenish than a threshold LCC shade (Singh, *et al.*, 2016).

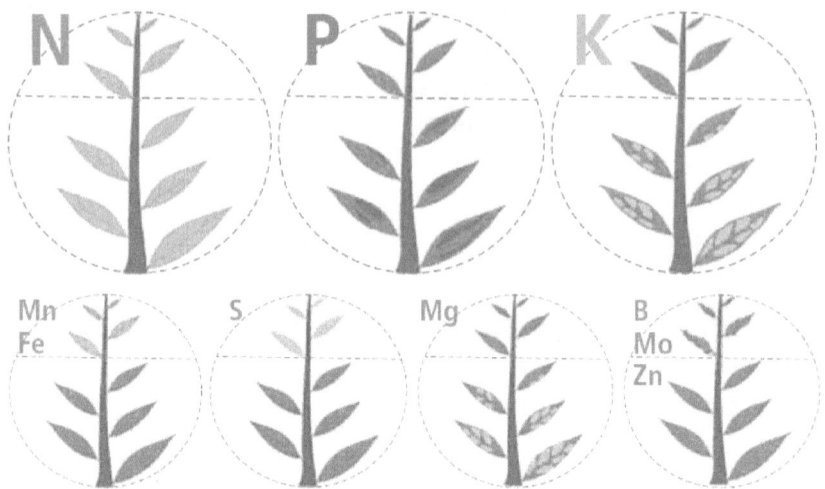

Fig 6.1: Common important nutrient deficiencies in plants

Nutrient	Location of Symptoms	Chlorosis	Leaf Margin Necrosis	Leaf Colour, Shape
N	All leaves	Yes	No	Yellowing of leaves, leaf veins
P	Older leaves	No	No	Purplish patches
K	Older leaves	Yes	Yes	Yellow patches
Mg	Older leaves	Yes	No	Yellow patches / Interveinal chlorosis
S	Young leaves	Yes	No	Yellow patches
Mn, Fe	Young leaves	Yes	No	Interveinal Chlorosis
B, Zn, Mo	Young leaves	-	-	Deformed Leaves

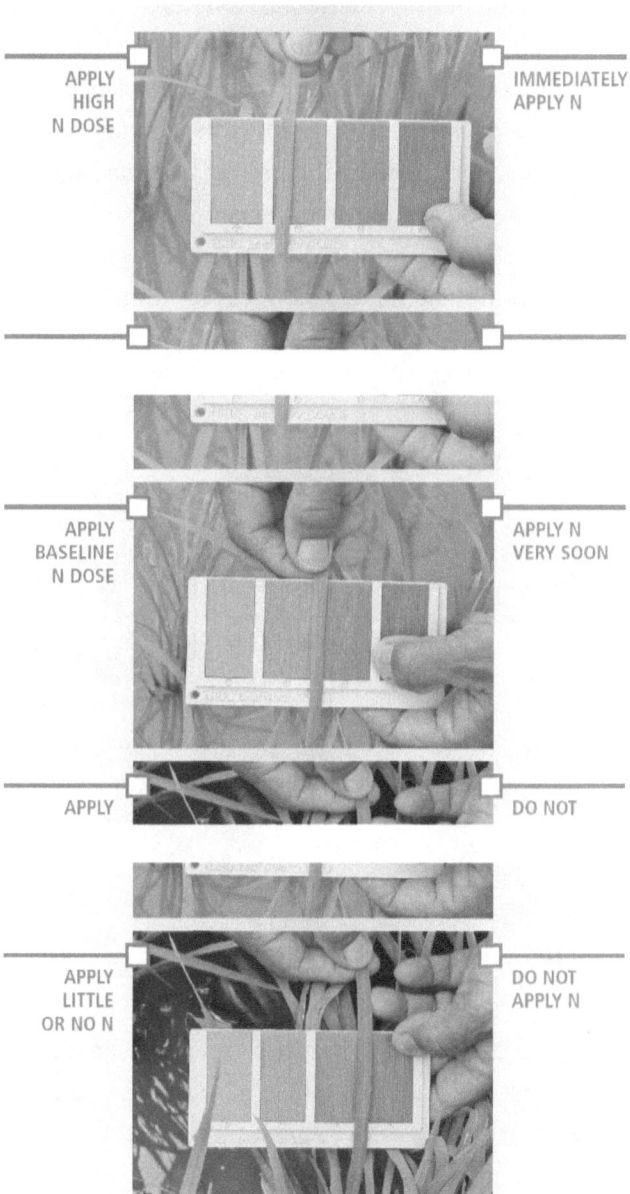

Fig. 6.2: *Nitrogen leaf color chart (LCC)*

SENSOR SYSTEMS

Site-specific nutrient management (SSNM) is a systems approach to management which involves decisions by the farmer and all of his input suppliers and advisers, each contributing his experience and training to the process. Extension workers, crop advisors, and farmers require easy-to-use tools that enable rapid identification of best management practices for specific rice-growing conditions. Decision support softwares, sensor systems and the LCC are among the tools that now help farmers pursue and determine best management practices based on SSNM.

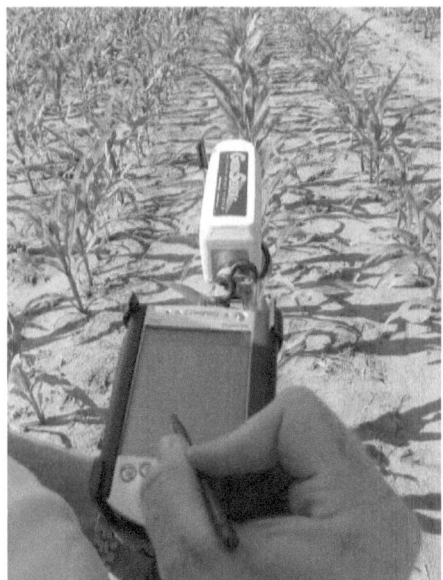

Fig. 6.3a: Hand held sensor

Fig. 6.3b: Machine mounted canopy sensor

New technologies are making various types of sensors available for determining status of some nutrients. These may be hand-held sensors, in-field plant exudate monitors, or machine-mounted canopy sensors.

Remote sensing imagery using scanners mounted on aircraft or satellites has been done, and more recently sensors mounted on unmanned aircraft (miniature airplanes or helicopters) have been used for in-field sensing of nutrient status. All of these tools depend upon having a good spectral signature of some plant nutrient response and a good set of calibration data to use in interpretation of the imagery.

SPAD (Soil Plant Analysis Development) Chlorophyll meter:

To precisely estimate leaf color, new sensor technologies have come into play in the form of a SPAD chlorophyll meter. The SPAD chlorophyll meter has been used since the 1990s by researchers and crop consultants to help estimate the N status of plants. It instantly measures chlorophyll content or "greenness" of plants to reduce the risk of yield-limiting deficiencies or costly over-fertilizing. The SPAD quantifies subtle changes or trends in plant health long before these are visible to the human eye. Non-invasive, non-destructive measurement is made on green plants, by simply clamping the meter over leafy tissue, and receiving an indexed chlorophyll content reading in less than 2 seconds.

Fig 6.4: The SPAD chlorophyll meter (Spectrum Technologies).

Thus SPAD is used to assess N needs by comparing in-field SPAD readings to university guidelines or to adequately fertilized reference strips. Research shows a strong correlation between SPAD measurements and leaf N content.

The SPAD meter can provide an indication of the N status of plants and then be used to manage fertilizer N in rice, wheat and maize on the lines as explained in the case of the LCC. However, unlike the LCC the SPAD mater can guide fertilizer N applications to crops when a sufficiency index (defined as SPAD value of the plot in question divided by that of a well-fertilized reference plot or strip) falls below 0.90 in rice or 0.95 in maize. This approach has the advantage that a critical SPAD value need not be worked out for different cultivars, climates or regions.

Other electronic tools that have become important for N management guidance include the GreenSeeker, the CropCircle, and the RapidScan CS-45 sensors. Commonly used as a single handheld unit, or mounted on a tool bar as a gang of multi-row sensors, these tools emit standard wavelength light beams and measure the reflected light coming back to the unit from the leaves. Recently, smaller handheld versions of these tools have become available.

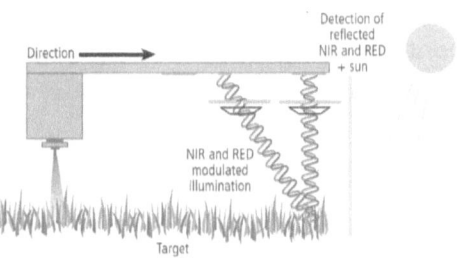

Fig. 6.5a: The Green Seeker system in use in the field.   *Fig. 6.5b: Operation of the GreenSeeker on-the-go sensing system*

The GreenSeeker sensor, emits light in two wavelengths, and then measures the reflectance from the crop canopy, and computes the NDVI value (Normalized Differential Vegetation Index) that relates to the amount of plant material in the field of view and its general vigor. The NDVI value is then compared to a calibration dataset, such as an N rate comparison strip, to provide a relative indication of plant condition that can be used to predict response to additional N fertilizer. The GreenSeeker may be used as a stand-alone, hand-held system for small areas or crop scouting, or a bank of multiple sensors may be mounted on a tractor or sprayer system and used for mapping or real-time variable-rate fertilizer application.

By calibrating with standard color references and "non-limited" N reference plots, an estimate of the N status of plants can be made and used to predict potential response to added N fertilizer. This is a variation on the color chart concept, but with the added feature of potentially geo-referencing the measurement, electronically storing the information, and wirelessly transmitting the results via the mobile telephone network. As cellular phone networks spread across rural areas throughout the world, tools such as this can potentially be used to improve N management wherever crops are grown.

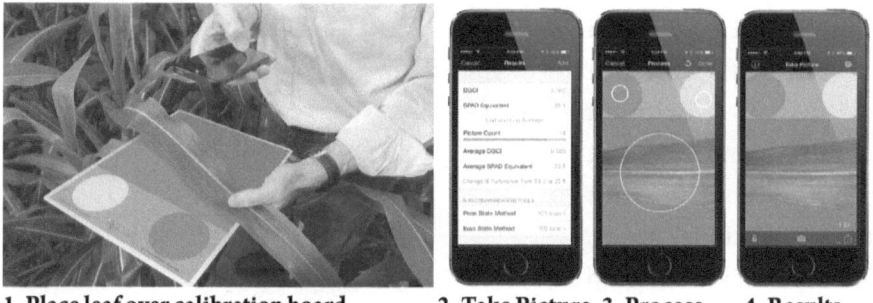

1. Place leaf over calibration board   2. Take Picture 3. Process   4. Results

Fig. 6.6: Field Scout Green Index + Nitrogen App
and Board (Spectrum Technologies).

A less expensive technology to aid in N management is based upon leaf images collected using a mobile "smart-phone" (Android or Apple) camera and reference board to scan plants in the field and assess their N status. The FieldScout GreenIndex application on the smart-phone interprets the "greenness" of the leaf, which can be used with calibration data to estimate N status of the plant. Recommendations for whether additional N fertilizer would be beneficial can then be made from the results.

With the GreenIndex system, a leaf is placed over the reference board, and photographed with a smart-phone. An area of the leaf photo is selected to compare with the reference colors. Then the images are processed by the phone software application, and results are presented based upon the recommendation algorithms stored in the smart-phone application. It is a very quick and simple procedure. Calibrations and recommendations can be developed for different crops and recommendation databases.

6.2 b Remote Sensing

In addition to in-field sensors to monitor crops, there are also *remote sensing* systems that use airplanes, satellites, and various kinds of drones to monitor crop conditions.

Remote sensing is the art and science of gathering information about the objects or area of the real world at a distance without coming into direct physical contact with the object under study. The principle behind remote sensing is the use of electromagnetic spectrum (visible, infrared and microwaves) for assessing the earth's features. The typical responses of the targets to these wavelength regions are different, so that they are used for distinguishing the vegetation, bare soil, water and other similar features. It can also be used in crop growth monitoring, land use pattern and land cover changes, water resources mapping and water status under field condition, monitoring of diseases and pest infestation, forecasting of harvest date and yield estimation, precision farming and weather forecasting purposes along with field observations. (Shanmugapriya, *et.al.*,2019).

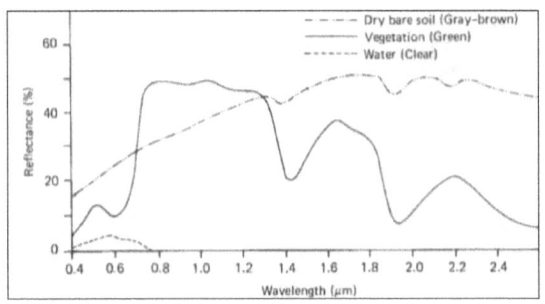

***Fig. 6.7**: Typical Spectral Reflectance curves for vegetation, dry bare soil and water*

Remote sensing data can greatly contribute to the monitoring of earth's surface features by providing timely, synoptic, cost-efficient and repetitive information about the earth's surface (Justice *et al.*, 2002). Remote sensing inputs combined with crop simulation models are very useful in crop yield forecasting. Since the ground based and air based platforms are time consuming and have limited use, these space based satellite technologies are gaining more importance for acquiring spatio-temporal meteorological and crop status information for complementing the traditional methods (Shanmugapriya, et.al.,2019).

AGRICULTURAL APPLICATIONS – BASIC ASPECTS

Remote sensing technology has the potential of revolutionizing the detection and characterization of agricultural productivity based on biophysical attributes of crops and/or soils (Liaghat and Balasundram, 2010). Data recorded by remote sensing satellites can be used for yield estimation (Doraiswamy *et al.*, 2005; Bernerdes *et al.*, 2012), crop phenological information (Sakamoto *et al.*, 2005), detection of stress situations (Gu *et al.*, 2007) and disturbances. Remote sensing has long been used in monitoring and analyzing of agricultural activities and sensing of agricultural canopies has provided valuable insights into various agronomic parameters. The advantage of remote sensing is its ability to provide repeated information without destructive sampling of the crop, which can be used for providing valuable information for precision agricultural applications. Remote sensing along with GIS is highly beneficial for creating spatio-temporal basic informative layers which can be successfully applied to diverse fields including flood plain mapping, hydrological modelling, surface energy flux, urban development, land use changes, crop growth monitoring and stress detection (Kingra *et al.*, 2016). In India, the satellite remote sensing is mainly used for the crop acreage and production estimation of agricultural crops.

The advances in the use of remote sensing methods are due to the introduction of narrow band or hyperspectral sensors and increased spatial resolution of aircraft or satellite mounted sensors. Hyperspectral remote sensing has also helped to enhance more detailed analysis of crop classification. Thenkabail *et al.*, (2004) performed rigorous analysis of hyperspectral sensors (from 400 to 2500 nm) for crop classification based on data mining techniques consisting of principal components analysis, lambda–lambda models, stepwise discriminant analysis and derivative greenness vegetation indices. Many investigations have included different types of sensors which are capable of providing the reliable data on a timely basis on a fraction of the cost of traditional method of data gathering.

MONITORING OF VEGETATION COVER

The science of remote sensing play a vital role in the area of crop classification, crop acreage estimation and yield assessment. Many research experiments were done using aerial photographs and digital image processing techniques. In relative to the crop condition, some remote sensing techniques are more focused on physical parameters of the crop system such as nutrient stress and water availability in assessing the crop health and yield.

Color and infrared aerial photography is used to photograph fields at critical growth stages for nutrition deficiency. The photos are then analyzed to determine areas which appear to be N-deficient. The photos geo-referenced along with soil maps, yield maps, and other information can be analyzed with GIS tools to determine possible points for additional scouting, sampling, or for N application. A quantitative relationship between relative greenness and yield loss is used to convert the aerial photograph into the yield loss map. This information

Fig. 6.8 i: Aerial Photo

about yield loss could be used to assess the economic impact of N deficiency, in support of a decision about whether to apply additional N fertilizer. The figure on the right shows yield loss (relative to the yield of the darkest areas in the aerial photograph) estimates derived from yield monitor data. The level of agreement between the predicted and observed yield loss maps suggests that remote sensing of N stress can provide a sound basis for making decisions about rescue N applications. Missing points in the predicted yield loss map are due to low certainty of canopy cover based on spectral properties.

Fig. 6.8 ii: Yield loss map based on yield monitor data

Fig. 6.8 iii: Yield loss map predicted from the aerial photo

Fig 6.8 i, ii and iii. *Processed images based upon aerial photography showing the N deficient areas identified in the early season aerial photograph accurately predicted yield losses*

Researchers are focused more on synoptic perspectives of regional crop condition using remote sensing indices. The most commonly used index to assess the vegetation condition is the Normalized Difference Vegetation Index proposed by Rouse *et al.*, (1974). The NDVI has become the most commonly used vegetation index (Calvao and Palmeirim, 2004, Wallace et al., 2004) and many efforts have been made aiming to develop further indices that can reduce the impact of the soil background and atmosphere on the results of spectral measurements.

The normalized difference vegetation index (NDVI), vegetation condition index (VCI), leaf area index (LAI), General Yield Unified Reference Index (GYURI), and Temperature Crop Index (TCI) are all examples of indices that have been used for mapping and monitoring drought and assessment of vegetation health and productivity (Doraiswamy et al., 2005, Ferencz et al., 2004, Prasad et al., 2006).

CROP CONDITION ASSESSMENT

Remote sensing can play an important role in agriculture by providing timely spectral information which can be used for assessing the biophysical indicators of plant health. The physiological changes that occur in a plant due to stress may change the spectral reflectance/emission characteristics resulting in the detection of stress amenable to remote sensing techniques (Menon, 2012). Crop monitoring at regular intervals of crop growth is necessary to take appropriate measures and also to know the probable loss of production due to any stress factor. The crop growth stages and its development are influenced by a variety of factors such as available soil moisture, date of planting, air temperature, day length, and soil condition. These factors are responsible for the plant conditions and their productivity. For example, corn crop yields can be negatively impacted if temperatures are too high at the time of pollination. For this reason, knowing the temperature at the time of corn pollination could help forecasters better predict corn yields (Nellis et al., 2009). The drought monitoring through satellite based information have been accepted in recent years and the use of Normalized Difference Vegetation Index (NDVI) and Vegetation Condition Index (VCI) have been accepted globally for identifying agricultural drought in different regions with varying ecological conditions (Nicholson and Farrar, 1994; Kogan, 1995; Seiler et al., 2000; Wang et al., 2001; Anyamba et al., 2001; Ji and Peters, 2003). Crop growth and its condition are often characterized through the use of various vegetation indices such as reflectance ratio, NDVI, PVI, transformed vegetation index, and greenness index. Annual NDVI profiles are extracted in operational remote sensing for

Vegetation Phenology Metrics (VPMs), and these metrics are used to characterize agricultural vegetation response to varying climatic and land management practices.

NUTRIENT AND WATER STATUS

The most important fields where we can opt for application of remote sensing and GIS through the application of precision farming are nutrient and water stress management. Detecting nutrient stresses by using remote sensing and GIS helps us in site specific nutrient management through which we can reduce the cost of cultivation as well as increase the fertilizer use efficiency for the crops. In semi-arid and arid regions judicious use of water can be made possible through the application of precision farming technologies. The advent of microwave remote sensing has made possible for estimating the soil moisture availability in the field. Information on crop water demand, water use, soil moisture condition, related crop growth at different stages can be obtained through the use of remote sensing data. With the increase in the development of hyper spectral bands in the thermal region, remote sensing has been playing a major role in understanding the crop soil characteristics. Such information when linked with GPS will provide promising results which are more helpful in precision farming (Shanmugapriya *et.al.*, 2019).

CROP EVAPO-TRANSPIRATION

The decline in the productivity of crops is due to irregularities in rainfall, increase in the temperature rate etc., which causes a decrease in the soil moisture. Drought is a situation which can be defined as a long-term average condition of the balance between precipitation and evapo-transpiration in a particular area, which also depends on the timely onset of monsoon as well as its potency (Wilhite and Glantz, 1985). In turn, vegetation indices such as CWSI (Crop Water Stress Index) (Jackson *et al.*, 1981), ST (Surface Temperature) (Jackson 1986), WDI (Water Deficit Index) (Moran *et al.*, 1994), and SI (Stress Index) (Vidal *et al.*,

1994) describe the relationship existing between water stress and thermal characteristics of plants.

Estimation of evapo-transpiration is essential for assessing the irrigation scheduling, water and energy balance computations, determining crop water stress index (CWSI), climatological and meteorological purposes. The energy emitted from cropped area has been useful in assessing the crop water stress as the temperature of the plants are mediated by the soil water availability and crop evapo-transpiration. Most of the approaches use simple direct correlations between remote sensed digital data and evapo-transpiration, but some combine various forms of remotely sensed data types. Remote sensing is playing a major role in the water management for agricultural system. And this can be further enhanced by the development of hyper spectral sensors and linking the remote sensing data with other spatial data through GIS and GPS technologies (Shanmugapriya et.al., 2019).

WEED IDENTIFICATION AND MANAGEMENT

Precision weed management technique helps in carrying out better weed management practices. Remote sensing coupled with precision agriculture is a promising technology nowadays. Though, ground surveying methods for mapping site–specific information about weeds are very time–consuming and labor–intensive. However, image–based remote sensing has potential applications in weed detection for site–specific weed management (Johnson et al., 1997; Moran et al., 1997; Lamb et al., 1999).

Based on the difference in the spectral reflectance properties between weeds and crop, remote sensing technology provides a mean for identifying the weeds in the crop stand and further helps in the development of weed maps in the field so that site specific and need based herbicide can be applied for the management of weeds. Weed prescription maps can be prepared with Geographic Information System (GIS), on the basis of which farmers can be advised to take the preventive control measures.

PEST AND DISEASE INFESTATION

Remote sensing has become an essential tool for monitoring and quantifying crop stress due to biotic and abiotic factors. Remote sensing methodologies need to be perfected for identification of insect breeding grounds for developing strategies to prevent their spread and taking effective control measures. The remote sensing approach in assessing and monitoring insect defoliation has been used to relate differences in spectral responses to chlorosis, yellowing of leaves and foliage reduction over a given time period assuming that these differences can be correlated, classified and interpreted (Franklin, 2001). The range of remote sensing applications has included detecting and mapping defoliation, characterization of pattern disturbances etc. and providing data to pest management decision support system (Lee *et al.*, 2010). Riedell *et al.*, (2004) reported remote sensing technology as an effective and inexpensive method to identify pest infested and diseased plants. They used remote sensing techniques to detect specific insect pests and to distinguish between insect and disease damage on oat. They suggested that canopy characteristics and spectral reflectance differences between insect infestation damage and disease infection damage can be measured in oat crop canopies by remote sensing.

CROP YIELD AND PRODUCTION FORECASTING

Remote sensing has been used to forecast crop yields primarily based upon statistical–empirical relationships between yield and vegetation indices (Thenkabail *et al.*, 2002, Casa and Jones 2005). The information on production of crops before the harvest is important for national food policy planning. Reliable crop yield is an important component of crop production forecasting purpose. The crop yield is dependent on many factors such as crop variety, water and nutrient status of field, influence by weeds, pest and disease infestation, weather parameters. The spectral response curve is dependent on these factors. The growth and decay in the spectral response curve indicates the crop condition and its performance.

PRECISION AGRICULTURE

Remote sensing technology is a key component of precision farming and is being used by an increasing number of scientists, engineers and large-scale crop growers (Liaghat and Balasundram, 2010). The main aim of precision farming is reduced cost of cultivation, improved control and improved resource use efficiency with the help of information received by the sensors fitted in the farm machineries. Variable rate technology (VRT) is the most advanced component of precision farming. Sensors are mounted on the moving farm machineries containing a computer which provides input recommendation maps and thereby controls the application of inputs based on the information received from GPS receiver (NRC, 1997). The advantage of precision farming is the acquisition of information on crops at temporal frequency and spatial resolution required for making management decisions.

ATMOSPHERIC DYNAMICS

Among the other applications through remote sensing, meteorological satellites are playing an important role in the forecasting of weather conditions. Meteorological satellites are designed to measure the atmospheric temperature, wind, moisture and cloud cover. The variations in the canopy temperature could indicate the areas of adequate and inadequate water in the field condition. The canopy temperature variability (CTV) is used in irrigation management and canopy air temperature difference (CATD) might be used as an indicator of crop water stress (Menon, 2012). Drought assessment playing a major role in the field of agriculture, wherein remote sensing data has been used for taking management decisions.

Sensing and communication tools are opening new possibilities for real-time monitoring and interpretation of plant nutrient status. Some of the technology is limited to large-area, intensively managed production systems. But an increasing amount of the technology is equally applicable to small-farm and economically stressed farming systems and provides modern tools that can benefit all farmers throughout the world.

To effectively utilize the information on crops for improvement of economy there is a need to develop state or district level information system based on available information on various crops derived from remote sensing and GIS approaches. The governments can use remote sensing data in order to make important decisions about the policies they will adopt or how to tackle national issues regarding agriculture. A new and nontraditional remote sensing application involves the implanting of nano-chips in plant and seed tissue that can be used in near-real time to monitor crop. Clearly, these and other new approaches will reinforce the importance of remote sensing in future analysis of agricultural sciences.

6.3 APPLICATION OF DRONE TECHNOLOGY IN AGRICULTURE

Drone technology provides enormous benefits and opportunities in a vast range of disciplines. Drones support tasks such as surveying, humanitarian work, disaster risk management, research, and transportation (Ayamga et al., 2020). In agriculture, drones can provide real-time imagery and sensor data from farm fields which cannot be quickly accessed on foot or by a vehicle (Malveaux *et al.*, 2014).

Drone or an unmanned aerial vehicle (UAV) according to the definition by International Civil Aviation Organization (ICAO, 2011) is an aircraft operated without a human pilot on-board. Another term, Unmanned Aircraft System (UAS) is defined as been made up of components such as a drone, the controller (ground-based), and the communication system between the two (the drone and the controller). Wallace *et al.* (2018) define two types of drones: (i) a fixed-wing – which generates lifts as it moves enabling it to sustain velocity through the air and (ii) the rotor which is highly manoeuvrable and can hover and rotate with a flight controller. These two types of drones have their advantages and disadvantages when it comes to flight range (endurance), battery capacity and payload (Ayamga *et al.*, 2021). Drone technology and the use of Unmanned Aerial Vehicles (UAVs) in such applications has brought about incremental progress, saving time and cost.

AGRICULTURAL DRONES

In recent years, the agricultural sector has seen some innovative technologies that support farm management strategies in enhancing efficiency through the precision application of farm inputs. One of such innovations is the drone technology which has gained popularity (Kim *et al.*, 2018) and has been widely used in precision agriculture (Zhang and Kovacs, 2012). As a part of the agricultural industry, drones are being employed for various operations in aerial surveillance, mapping, land inspection, monitoring, spraying fertilizers, checking for diseased or rotting crops, and much more. or crop fertilization, drones such as quadcopters prove to be the most favorable owing to their multi-rotors. Fixed-wing drones suit the purpose of crop fertilization, albeit their large structure requiring a large space for take-off and landing comes in the way.

FERTILIZER SPRAYING

One of the most critical uses of drones in agriculture is its flexibility to move around in swift motions and maneuver to the destined locations. This ability of drones helps spray fertilizers and insecticides to nurture

Fig 6.9: *Use of drones for spraying nutrients and pesticides*

crops and provide them with the needed nutrients. Such reinforcements allow crops to be healthy and flourishing. The drone operators are free to monitor the drone spraying fertilizers that keep insects, pests, and worms away and increase crop life longevity.

SOIL HEALTH MONITORING

Agricultural drones are providing real-time data that enables farmers to make informed decisions regarding the use of farm inputs. The robust capabilities of drones help in the primitive operation of analyzing soil health. In essence, UAVs collect and process data received from monitoring that can help check, control, and maintain the soil's health and nature. Drone technology can also provide the essential nutrients to the soil to improve their health and well-being. Through its operations of 3D mapping and data processing, drones achieve this operation of analyzing soil health.

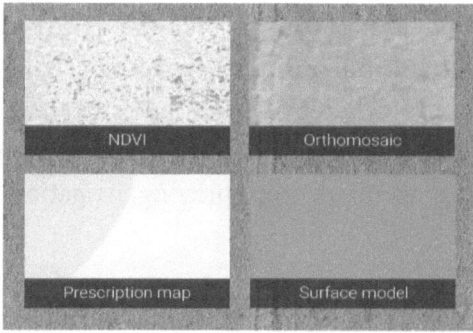

Fig 6.10: Use of drones for field imaging

SEEDING PROCESS

By its virtue, agriculture is a highly cumbersome and exhaustive industry because it requires skillful abilities to perform its operations. Seeding, especially, requires manual labor as it is a time-consuming procedure. To ease this tiring process, drone technology is employed to sow the seeds of the copious varieties of crops. As instilled in drones, the lasers, sensors, tanks, etc., allow them to quickly yet smoothly plant seeds.

ANALYZING DEFICIENCIES

Another incredible merit of adopting drones for crop fertilization comes with their feature to analyze, identify, and survey the crops for any deficiencies. Their high-resolution cameras and sensors, additionally instilled with lasers, help to perform these operations quickly. Unmanned Aerial Vehicles can also map these deficiencies in real-time, and the data collected and processed can be used to make further determinations regarding the crops.

The use of drone technology in agriculture is currently helping agricultural businesses meet the changing and growing demands of the future, whereby drones helps to increase efficiency in certain aspects of the farming process, from crop monitoring to planting, livestock management, crop spraying, irrigation mapping, and more.

AGRICULTURE DRONE- Future of farm activities

A DRONE (Dynamically Remotely Operated Navigation Equipment), in technological terms, is an unmanned aircraft. Drones are more formally known as unmanned aerial vehicles (UAV's) or unmanned aircraft systems (UAS's). Essentially, a drone is a flying robot that can be remotely controlled or fly through software.

The British Oxford Dictionary defines the word DRONE as a continuous low noise and this low noise later got a meaning of an aircraft controlled from ground.

The first pilotless radio-controlled aircraft were used in World War I. In 1918, the U.S. Army developed the experimental Kettering Bug, an unmanned "flying bomb" aircraft, which was never used in combat.

To achieve flight, drones consist of a power source, such as battery or fuel, rotors, propellers and a frame. The frame of a drone is typically made of lightweight, composite materials, to reduce weight and increase maneuverability during flight. Drones require a controller, which is used remotely by an operator to launch, navigate and land it. Controllers communicate with the drone using radio waves.

What is new in drone technology:

Forest Fire Protection with Drones: During forest fire or wildfire it is extreme dangerous for human to extinguish fire, where these drones can deploy autonomously and extinguish the fire of 50 cubic meters of fire extinguishing area per forest.

Search and Rescue Drones: Through the megaphone system mounted in the drone, voice prompts or notifications can be given for the people to react and required materials can be dropped to rescue the affected people.

Agriculture drone: *An agricultural drone is an unmanned aerial vehicle used to help optimize agriculture operations, increase crop production, and monitor crop growth. Sensors and digital imaging capabilities can give farmers a richer picture of their fields. Using an agriculture drone and gathering information from it may prove useful in improving crop yields and farm efficiency.*

Aerial Mapping: *Geospatial technology is what gives UAS the ability to be autonomous. The convenience of inspecting vast infrastructure without significant time and manpower invested is enough of a reason for surveyors, construction firms, and power companies to deploy drones.*

Beneficiary to Agriculture farmers:

The specially designed agriculture sprayer drone's protect farmers from poisoning and heart stroke, while spraying liquid pesticides, fertilizers and herbicides on agriculture land. At least 20 acres can be sprayed per day per drone, which is 10 times more than the traditional knapsack sprayer and 80-90% of water can be saved in comparison to traditional spraying methods.

Comparison of different sprayers used in agriculture

	Mist Blower	Hydrogun	Boomer	Power Sprayer	Agriculture Drone
Droplet size (Micron)	50-100	200-400	100-150	-	50-200
Quantity (L/acre)	20-80	150	200	150	15-20
Efficiency (acres/day)	1.5-2	-	7-20	4	25-30
Pesticides utilization efficiency	30-40%	-	30%	30-40%	85%

Features of agriculture drone:

Sensors to detect to avoid obstacles: Drones can detect centimeter wide power lines from up to 15 m away. This protects the drone not only from power lines, but from trees branches and other common obstacles as well.

Active obstacle sensing and avoidance works during the day or at night without being influenced by light or dust.

Auto Operation mode: *User can select the operation area on APP, and set operation distance, flight speed, altitude and other information. The drone will automatically fly back and forth according to the specified distance and traverse the entire area to complete the work, and the land operation is more convenient.*

Terrain Following RADAR: *Drone can scan the terrain below it in real time to keep a constant, centimeter-accurate height above crops. Spray density is maintained even as the ground rises and drops so that an optimal amount of liquid is applied at all times.*

Regulation constraints to operate agricultural drone:

- *NPNT compliance is mandatory*
- *Operator must obtain security clearance from Ministry of Home Affairs (MHA)*
- *Permission artifact should be obtained from DGCA with 24 hours prior the operation*
- *Right from the micro-nutrition to insecticides or pesticides can be sprayed through the drone for the plants at each stage. So, permission should be taken from Central Insecticides Board (CIB) for aerial spraying*
- *Agricultural drone should obtain special clearance from DGCA*
- *Local police permission should be obtained before the drone operations*

Support required for farmers and company:

- *Train Rural Youth to fly UAV (Drones) under government schemes and Programs*
- *Establish Custom Hire Service Center for Agriculture Drones in panchayath level*

- *Subsidy to purchase Agriculture Drones*
- *Agricultural drone regulations should be converged for the ease of drone operations*

Shivakumar H.G & Aditya. A.R
Multiplex Drone Pvt Ltd, Bangalore
IMMA Souveneir, 2018

Agricultural Micronutrients- Industry Trends

7.1 AGRICULTURAL MICRONUTRIENTS- MARKET IN INDIA

The India agricultural micronutrients market segmentation is based on form, type, mode of application, crop type, regional distribution, top 3 states analysis in each region, and competitional landscape. Based on type, the market is further segmented into zinc, copper, boron, iron, manganese, and others. The depleting micronutrients levels are alarming and creating surge in the demand for the external supply of micronutrients thereby supporting the growth of the India agricultural micronutrients market in the next five years. The iron sub-segment is anticipated to register significant growth in the future five years.(*https:// www.techsciresearch.com/report/india-agricultural-micronutrients-market/7776.html*)

The India agricultural micronutrients market size was USD 538.4 million in 2021. The market is projected to grow from USD 571.6 million in 2022 to USD 1,057.6 million by 2029, exhibiting a CAGR of 9.19% during the forecast period. The global COVID-19 pandemic has been unprecedented and staggering, with experiencing lower-than-anticipated demand across all regions compared to pre-pandemic levels. Based on analysis, the global market exhibited a decline of 5.23% in 2020 as compared to 2019. (*https:// www.fortunebusinessinsights.com/india-agricultural-micronutrients-market-106899*).

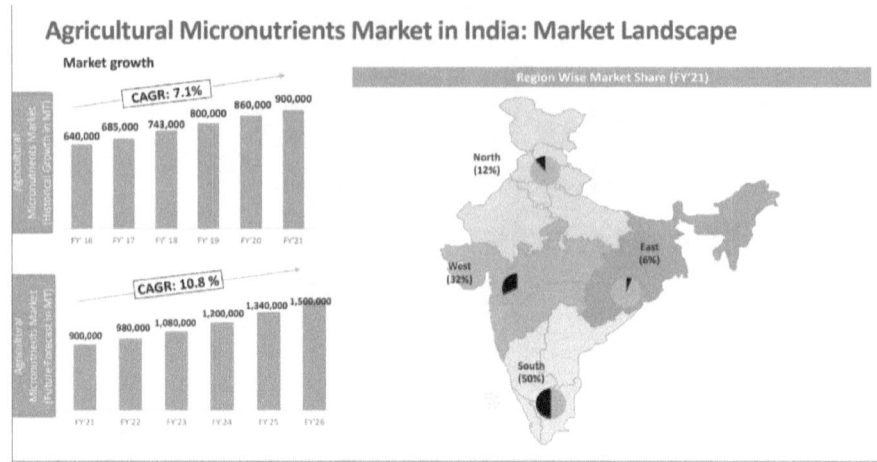

Fig. 7.1: *Agricultural Micronutrients Market in India- Market Landscape*
Source: V4C Research 2021: Agricultural Micronutrients Market in India (FY'21-26)

DRIVING FACTORS

Increasing Micronutrient Deficiencies in Agricultural Soils of India

In India, intensive farming and changes in environmental conditions, such as increase in global warming, are the major factors, which contributed to the depletion of micronutrients in soil. Hence, crop cultivation in micronutrient deficient soils has adversely affected the crop yield, thereby leading to crop losses. Additionally, lack of adoption of timely soil testing has hampered the sustainable production of agricultural crops in India.

Thus, fortification of fertilizers with essential micronutrients is an efficient solution to eliminate the aforementioned problem as various micronutrients are able to deal with a wide range of soil conditions/problems. Moreover, the increased demand for high-quality and fully-developed fruits, vegetables, other horticulture, and ornamental crops worldwide is projected to play a crucial role in augmenting the demand for micronutrient enriched fertilizers. Thus, micronutrient fertilizers are gaining widespread recognition as they are the best fit for modern sustainable agriculture.

Strong Demand for High-Value Crops and Government Support to Boost Growth

In recent years, the Indian consumer inclination has shifted toward healthy foods to maintain and improve health. The increasing demand for high-value crops such as nutrient-rich fruits and vegetables is primarily fuelling the demand for micronutrients across the Indian agriculture Industry. In addition, rising knowledge about the relation between healthy crops and increased yield has amplified the agricultural micronutrients market growth.

Furthermore, the Indian government is taking initiatives to educate farmers about soil health by launching lucrative agricultural schemes and offering subsidies to promote the adoption of agriculture micronutrients. For instance, in 2020, the Government of India (GOI) implemented the Nutrient Based Subsidy (NBS) scheme and promoted fortified and customized fertilizers. The government also offered additional subsidies on zinc and boron to promote their usage alongside primary nutrients.

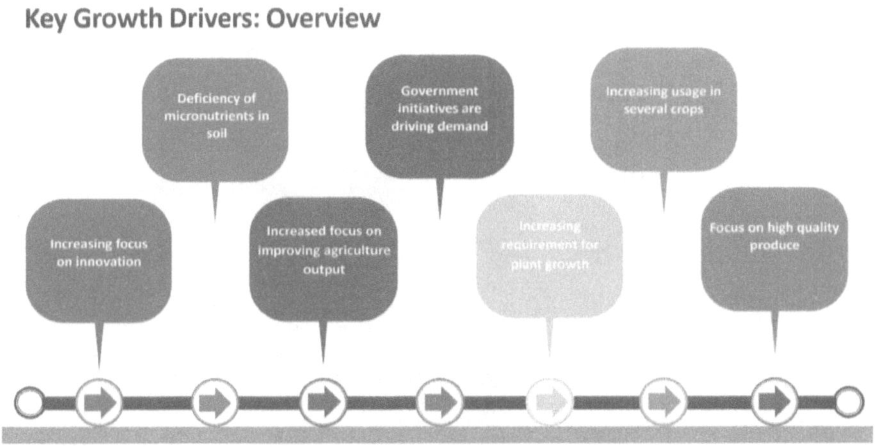

Key Growth Drivers: Overview

- Deficiency of micronutrients in soil
- Government initiatives are driving demand
- Increasing usage in several crops
- Increasing focus on innovation
- Increased focus on improving agriculture output
- Increasing requirement for plant growth
- Focus on high quality produce

Fig. 7.2: Key Growth Drivers- Overview
Source: V4C Research 2021: Agricultural Micronutrients Market in India (FY'21-26)

RESTRAINING FACTORS

Unregulated Application and Low Adoption of Micronutrients to Hamper Market Growth

The primary factor restraining the growth of the micronutrients market is low awareness regarding the benefits of these products among farmers in India. Lack of awareness regarding the nutritional status of their farm lands among low-income farmers in the country limits the adoption of micronutrients. The efficacy of micronutrient products significantly depending on the rate of application and mode of application. It is important to identify proper application rate and method of application. The excessive application of micronutrients can cause soil toxicity and negatively impact crop productivity. Thus, lack of proper knowledge regarding efficient nutrition (other than conventional fertilizers) required by crops to attain healthy and timely growth further hampers the agricultural micronutrients market growth.

AGRICULTURAL MICRONUTRIENT MARKET – BASED ON TYPE

Based on type, the market is divided into zinc, boron, iron, molybdenum, manganese, and others.

Mineral elements support growth of cereals, spices, oilseeds, pulses, and plantation. The unavailability of micronutrients causes deformations, reduced growth, and reduced yield. Micronutrients are essential for the growth of plant and play

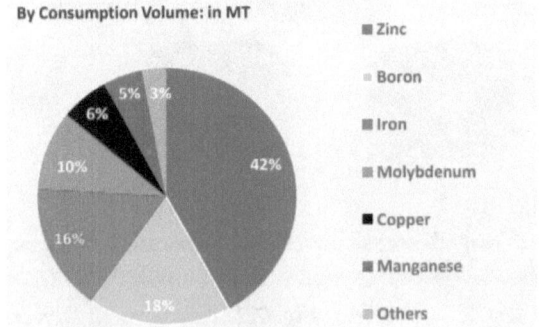

Fig. 7.3: Agricultural Micronutrient Market – Based On Type

Source: V4C Research 2021: Agricultural Micronutrients Market in India (FY'21-26)

a significant part in supporting crop nutrition. Zinc deficiency is one of the major concerns among Indian farmers. In addition, low hormone production due to zinc deficiency causes the shortening of internodes and stunted leaf growth in crops. Thus, zinc is required for protein synthesis, growth regulation, synthesis of plant-growth substances, and is essential for promoting certain metabolic reactions, which are particularly critical in the early growth stages.

Iron is a catalyst to chlorophyll formation, Iron acts as an oxygen carrier in the nodules of legume roots. The growing knowledge of the benefits of iron among farmers is surging the demand for ferrous fertilizers.

MICRONUTRIENT MARKET – BY SOURCE ANALYSIS

Based on source, the market segmented into chelated and non-chelated.

The increasing knowledge regarding advancing technologies among farmers is expected to amplify the use of micronutrient chelates. Chelation is an attachment of chelating agents to metal ions. Chelates are considered to be organic molecules. Chelation is important as it makes the metal ions more available for efficient uptake. Moreover, chelates are available for different ranges of soil pH, which increase their adoption among farmers.

Traditionally, non- chelated micronutrient products have been extensively used by farmers due to their wider availability and easy application. The lack of knowledge regarding chelated micronutrients among low income farmers has increased the value share of non-chelated micronutrients in the market.

MICRONUTRIENT MARKET – BY APPLICATION ANALYSIS

Based on application, the market is segmented into soil, foliar, and fertigation.

Soil application is the most effective method for micronutrient application due to its low cost and ease of application. Lack of advanced fertilization techniques in various parts of India and dearth of knowledge among farmers have driven the adoption of soil application method.

However, foliar application is expected to witness strong growth in the coming years due to its uniform application. Moreover, increasing investment by manufacturers for commercializing the foliar products is projected to fuel the adoption of foliar application in the foreseeable years.

MICRONUTRIENT MARKET – BY CROP TYPE ANALYSIS

Based on crop type, the market is segmented into cereals, pulses and oilseeds, fruits and vegetables, and others.

Cereal segment is projected to dominate the market attributed to the increasing prevalence of nutrient loss in cereal crops such as wheat, rice, and others. The growing demand for cereal crops as a staple food has significantly increased the demand for micronutrient fertilizers for these crops.

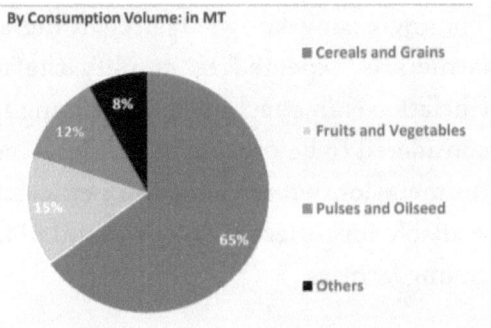

Fig. 7.4: Agricultural Micronutrient Market – By Crop Type Analysis
Source: V4C Research 2021: Agricultural Micronutrients Market in India (FY'21-26)

The demand for high-quality oilseeds & pulses and fruits & vegetables has significantly increased across the India market due to its nutritious value. This, in turn, has amplified the demand for these micronutrients for application on these crops to offer good quality crops in the domestic as well as international markets. This is one of the major reasons for the growth of the agricultural micronutrients market.

India is the second biggest consumer of fertilizer in the world next only to China. The demand for fertilizers in the country has been growing due to increasing demand for food grains on account of the country's burgeoning population. The increasing population growth has led to food security concern and increased awareness among the farmers about the uses of fertilizers. Some of the other factors driving the growth of the market are high government subsidies and growing investments in the fertilizer industry. Fertilizers have played a key role in the success of India's green revolution and subsequent self-reliance in food-grain production. The increase in fertilizer consumption has contributed significantly to sustainable production of food grains in the country. As a result, the demand of fertilizers has witnessed double digit growth rates over the past several years. Increasing rural incomes, coupled by easy availability of credit, are also likely to create a positive impact on fertilizer usage in the country.

Bibliography

Abdel-Aziz, H.M.M., Hasaneen, M.G.A. and Omer, A.M. (2016). Nano chitosan-NPK fertilizer enhances the growth and productivity of wheat plants grown in sandy soil. *Spanish Journal of Agricultural Research*, 14(1): 9.

Ahmed, S., Niger, F., Kabir, M.H., Chakrabarti, G., Nur, H.P. and Imamul Huq, S.M. (2012). Development of Slow Release Nano Fertilizer. In: Proceedings of the International Workshop on Nanotechnology, September 21-23, Dhaka, Bangladesh. ISBN No. 978-986-33-5715- 1.pp. 45.

Amanullah., Farooq, A., Rehman, M.S., Arshad, H., Shoukat, A., Nadeem, A., Nawaz, A. Wakeel, A. and Nadeem, F. (2018). Manganese nutrition improves the productivity and grain biofortification of bread wheat in alkaline calcareous soil. *Experimental Agriculture*, 54(5): 744–754.

Andersson, M., Karumbunathan, V. and Zimmermann, M.B. (2012). Global iodine status in 2011 and trends over the past decade, *Journal of Nutrition*, 142(4):744-750.

Anyamba, A., Tucker, C.J. and Eastman, J.R. (2001). NDVI anomaly patterns over Africa during the 1997/98 ENSO warm event. *International Journal of Remote Sensing*, 22(10): 1847–1859.

Asher, C. J., Reghenzani, J. R. Robards, K. H and D. E. Tribe (1990): Rare Earths in Chinese Agriculture. Australian Academy of Technological Sciences and Engineering, Victoria, Australia

Asher, C.J.(1991). Beneficial Elements, Functional Nutrients, and Possible New Essential Elements in *Micronutrients in Agriculture*, Volume 4, Ed; 2

Atiqueur, R., Farooq, M., Cheema, Z.A. and Wahid, A. (2012). Seed priming with boron improves growth and yield of fine grain aromatic rice. *Plant Growth Regulation*, 68(2)189–201.

Ayamga, M., Tekinerdogan, B., Kassahun, A. and Rambaldi, G.(2020). Developing a policy framework for adoption and management of drones for agriculture in Africa. *Technology Analysis and Strategic Management*, 33(8): 1–18.

Babar, N.F., Muzaffar, R., Khan, M.A. and Imdad, S. (2010). Impact of socioeconomic factors on nutritional status in Primary School Children. *Journal of Ayub Medical College, Abbottabad*, 22(4): 15-18.

Banfield, J.F. and Zhang, H. (2001). Nanoparticles in the Environment, *Rev. Min. Geochem.* 44(1):1-58.

Behera, S.K., Lakaria, B.L., Singh, M.V. and Somasundaram, J. (2011). Molybdenum in soils, crops and fertilizers: An overview. *Indian Journal of Fertilisers*, 7(5): 52-57.

Behera, S.K., Shukla, A.K. and Lakaria, B.L. (2014). Deficiency of boron and molybdenum in soils and crops in India and their amelioration. *Ind Farm*, 63(12):27-29.

Behera, S.K. and Shukla, A.K. (2014). Total and extractable manganese and iron in some cultivated acid soils of India - status, distribution and relationship with some soil properties. *Pedosphere* 24, 196- 208.

Bell, P. F., Chaney, R. L. and Angle, J. S. (1991). Free metal activity and total metal concentrations as indices of micronutrient availability to barley [*Hordeum vulgare* (L.) "Klages"]. In: Iron Nutrition and Interactions in Plants. pp. 69-80. Chen, Y. and Hadar, Y., Eds., Kluwer Academic Publishers, Dordrecht.

Benzon, H.R.L., MRU, R., Ultra, V. Jr. and Lee, S.C. (2015). Nano-fertilizer affects the growth, development and chemical properties of rice. *International Journal of Agronomy & Agricultural Research,* 7(1):105-117.

Bennett, W.F. (ed). 1993. Nutrient Deficiencies and Toxicities in Crop Plants. St. Paul, Minn. APS Press. 202 p. Brady and Weil, 1999.

Bernardes, T., Meriera, M. A., Adami, M., Giarolle, A. and Rudorff, B. F. T. (2012). Monitoring biennial bearing effect on coffee yield using MODIS remote sensing imagery. *Remote Sensing,* 4(9): 2492 – 2509.

Bhupalraj, G.; Patnaik, M.C.; Khadke, K.M. (2002). Molybdenum status in soils of Andhra Pradesh. *AICRP Micro Second. Nutr. Soils Plants Pradesh,* 36, 1–87.

Bockman, O. C., Kaarstad, O., Lie, O.H. and Richards, I. (1990). Agriculture and fertilizers: Fertilizers in perspective. Agricultural Group, Norsk Hydro, Oslo.

Boller, B. and Greene, S.L. (2010). Genetic resources. In *Fodder Crops and Amenity Grasses,* Springer: New York, NY, USA, pp. 13–37.

Bouis, H.E. and Saltzman, A. (2017). Improving nutrition through biofortification: A review of evidence from HarvestPlus, 2003 through 2016. *Global Food Security,* 12: 49–58

Brady, N.C. and Weil, R.R.(2005). The Nature and Properties of Soils. (13th edi.). Pearson Education Inc., New Jersey, USA. 559-592.

Brunnert, I., P. Wick, Manserp, Spohnp, R.N. Grass, L.K. Limbach, A. Bruinink and W.J. Stark (2006). In vitro cytotoxicity of oxide nanoparticles: comparison to asbestos, silica and effect of particle solubility.Environmental Science and Technology., 40: 4374-4381.

Buffle J. (2006). "The key role of environmental colloids/nanoparticles for the sustainability of life". *Env Chem.* 3(3): 155-158.

Cabral, G.B., Carneiro, V.T., Rossi, M.L., da Silva, J.P., Martinelli, A.P. and Dusi, D.M. (2015). Plant regeneration from embryogenic callus and cell suspensions of *Brachiaria brizantha*. *In Vitro Cellular & Developmental Biology - Plant*, 51(3):369–377.

Cakmak, I. and Kutman, U.B. (2017). Agronomic biofortification of cereals with zinc: A review. *European Journal of Soil Science*, 69: 172–180.

Calles, T., del Castello, R., Baratelli, M., Xipsiti, M. and Navarro, D.K.(2019). The International Year of Pulses; Final Report; FAO: Rome, Italy, 2019; p. 40.

Calvao, T. and Palmeirim, J.M. (2004). Mapping Mediterranean scrub with satellite imagery: biomass estimation and spectral behaviour. *International Journal of Remote Sensing*, 25(16):3113-3126

Caravalho, A., Reis, S., Pavia, S., Lima-Brito, I. and Eduardo, J. (2019). Influence of seed priming with iron and/or zinc in the nucleolar activity and protein content of bread wheat. *Protoplasma*, 256(3): 763–775.

Casa, R. and Jones, H.G. (2005). LAI retrieval from multiangular image classification and inversion of a ray tracing model. *Remote Sensing of Environment*, 98(4): 414–428.

Chinnamuthu, C. R. and Kokiladevi, E. (2007). Weed management through nanoherbicides. In Application of Nanotechnology in Agriculture, C. R. Chinnamuthu, B. Chandrasekaran and C. Ramasamy (Eds), Tamil Nadu Agricultural University, Coimbatore, India.

Combs, G.F., Welch, R.M., Duxbury, J.M., Uphoff, N.T. and Nesheim, M.C. (1996). Food- Based Approaches to Preventing Micronutrient Malnutrition: An International Research Agenda.: 1-68. Cornell International Institute for Food, Agriculture and Development. Cornell University, Ithaca, NY.

Dai, H., Wei, S. and Twardowska, I. (2020). Biofortification of soybean (*Glycine max* L.) with Se and Zn, and enhancing its physiological functions by spiking these elements to soil during flowering phase. *Science of the Total Environment*, 740: 139648.

Davis, G.K., and Mertz, W. (1987). Copper. In Trace Elements in Human and Animal Nutrition, 1(5): 301– 364 (W. Mertz, Ed.), Academic Press, San Diego, California, USA.

Dhaliwal, S.S. and Manchanda, J.S. (2009). Critical level of Boron in Typic Ustrochrepts for predicting response of mungbean (*Phaseolus aureus* L.) to boron application. Indian Journal of Ecology, 36 (1):22–27.

Dhaliwal, S.S., Sadana, U.S., Manchanda, J.S. and Dhadli, H.S. (2009). Biofortification of wheat grains with zinc (Zn) and iron (Fe) in typic Ustochrept soils of Punjab. *Indian Journal of Fertilizers*, 5(11):13–16.

Dhaliwal, S.S., Sadana, U.S., Khurana, M.P.S., Dhadli, H.S. and Manchanda, J.S. (2010). Enrichment of paddy grains (*Oryza sativa* L.) with zinc and iron through ferti-fortification. *Indian Journal of Fertilizers*, 6: 28–35.

Dhaliwal, S.S., Sadana, U.S. and Manchanda, J.S. (2011). Relevance and essentiality of ferti-fortification of wheat grains with manganese and copper. *Indian Journal of Fertilizers*, 7(11): 48–55.

Dhaliwal, S.S., Sadana, U.S., Khurana, M.P. and Sidhu, S.S. (2012). Enrichment of wheat grains with Zn through ferti-fortification. *Indian Journal of Fertilizers*, 8(7): 34-45.

Dhaliwal, S.S., Sadana, U.S., Manchanda, J.S., Khurana, M.P.S. and Shukla, A.K. (2013). Differential response of maize cultivars to iron (Fe) applied through ferti-fortification. *Indian Journal of Fertilizers*, 9: 52–57.

Dhaliwal, S.S., Sandhu, A.S., Shukla, A.K., Sharma, V., Kumar, B. and Singh, R. (2020). Bio-fortification of oats fodder through zinc enrichment to reduce animal malnutrition. *Journal of Agricultural Science and Technology A,* 10 (2): 98–108.

Dhaliwal, S.S., Sharma, V., Shukla, A.K., Verma, V., Kaur, M., Shivay, Y.S., Nisar, S., Gaber, A., Brestic, M., Barek, V., Skalicky, M., Ondrisik, P. and Hossain, A. (2022). Biofortification—A Frontier Novel Approach to Enrich Micronutrients in Field Crops to Encounter the Nutritional Security, *Molecules,* 27(4), 1340.

Dhillon, K.S. and Dhillon, S.K. (1991). Selenium toxicity in soils, plants and animals in some parts of Punjab, India. *Int J Env St* 37(1-2): 15-24.

Doraiswamy, P. C., Sinclair, T. R., Hollinger, S., Akhmedov, B., Stern, A. and Prueger, J. (2005). Application of MODIS derived parameters for regional crop yield assessment. *Remote Sensing of Environment,* 97(2): 191 – 202.

Edejer, T. T. T., Aikins, M., Black, R., Wolfson, L., Hutubessy, R., & Evans, D. B. (2005). Cost effectiveness analysis of strategies for child health in developing countries. *Bmj,* 331(7526), 1177.

Eskew, D. L., Welch, R. M. and Norvell, W. A. (1983). Nickel: an essential micronutrient for legumes and possibly all higher plants. *Science,* 222:621-623.

Eskew, D. L., Welch, R. M. and Norvell, W. A. (1984). Nickel in higher plants further evidence for an essential role. *Plant Physiology,* 76:691-693.

Fageria, N.K., Baligar, V.C. and Jones, C.A. (1997). Growth and mineral nutrition of field crops. 2nd edition, Dekker, New York.

Fan, S. and Saurkar, A. (2006). Public spending in developing countries: trends, determination and impact (mimeo). FAO (Food and Agriculture

Organisation of the United Nations), FAOSTAT online databases, Rome (http://faostat.fao.org/default.aspx).

Fan, S., Gulati, A. and Thorat, S. (2008). Investment, subsidies, and pro-poor growth in rural India. *Agricultural Economics* 39(2):163-170.

Farooq, M., Basra, S.M.A., Tabassum, R. and Afzal, I. (2006). Enhancing the performance of direct seeded fine rice by seed priming. *Plant Production Science,* 9(4):446–456.

Ferencz., C. S., P., Bognar, J., Lichtenberger, D., Hamar., G. Y. Tarcsai, G., Timar, G., Molnar, S.Z., Pasztor, P., Steinbach, B., Szekely, O. E., Ferencz. and Ferencz- Arkos, I. (2004). Crop yield estimation by satellite remote sensing. *International Journal of Remote Sensing*, 25(20): 4113– 4149.

Fleischer, M. and Ehwald, O.R. (2014). The pore size of non-graminaceous plant cell wall is rapidly decreased by borate ester cross-linking of the pectic polysaccharide rhamnogalacturonan II. *Plant Physiology.* 121(3):829-838.

Franklin, S. (2001). Remote Sensing for Sustainable Forest Management. Lewis publisher, Boca Raton, Florida, p.407.

Fujita, T., Takahashi, C., Ushioda, T. and Shimizu, H. (1983). Coated granular fertilizer capable of controlling the effects of temperature upon dissolution-out rate. United States Patent, No. 4, 881,963.

Fujita, T. and Shoji, S. (1999). Kinds and properties of Meister fertilizers. In: Meister controlled release fertilizer – Properties and Utilization. Shoji, S. (ed). Konno Printing Company Ltd. Sendai, Japan. pp. 13-34.

Ganeshamurthy, A.N. and Raghupathi, H.B.(2014). Soil and plant testing in nutrient management in perennial horticultural crops. In Soil Testing for Balanced Fertilisation: Technology-application-problems solutions (H.L.S. Tandon, Ed.), pp. 68- 88. FDCO, New Delhi.

Goertz, H.M. (1993). Controlled Release Technology. Kirk-Othmer Encyclopedia of Chemical Technology, Vol.7 Controlled Release Technology (Agricultural), pp. 251- 274.

Grewal, J.S. and Trehan, S.P. (1990) Micronutrient deficiency and amelioration in potato. Shimla Tech. Bull., CPRI, 23: 1–26.

Gu, Y., Brown, J. F., Verdin, J. P. and Wardlow, B. (2007). A five year analysis of MODIS NDVI and NDWI for grassland drought assessment over the central great plains of the United States. *Geophysical Research Letters*, 34(6):34.

Gulati, A., Sharma, P., Samantra, A. and Terway, P. (2018). Agriculture Extension System in India- Review of Current Status, Trends and the Way Forward. Report for Indian Council For Research On International Economic Relations (ICRIER). e-ISBN: 978-81-937769-0-2

Gupta, G., Dhar, S., Kumar, A., Jinger, D., Kumar, V., Kumar, A. and Kamboj, N.K. (2018). Specialty Fertilizers: Status, Prospects and Significance In India, *Adv.Agr and Biodiv.*

Hähndel, R. (1986). Slow-release Fertilizers - their Properties and Advantages. 4/1986. BASF Germany.

Hassan, S.R., Singh, B and Gaffar, M.A. (1985). Indian Journal of Veterinary Medicine 5, 132-134.

Hegde, D. M., and S. N. Sudhakar Babu. 2009. Declining factor productivity and improving nutrient-use efficiency in oilseeds. Indian Journal of Agronomy 54 (1):1–8.

Hetzel, B.S. and Pandav, C.S.(1994). S.O.S. for a Billion. The Conquest of Iodine Deficiency Disorders. Oxford, Oxford University Press.

Hope, B. (1997). An assessment of the global impact of anthropogenic vanadium. *Biogeochem*, 37: 1-13.

Huang, S., Wang, P., Yamaji, N. and Ma, J.F.(2020). Plant nutrition for human nutrition: Hints from rice research and future perspectives. *Mol. Plant*, 13(6): 825–835.

Hunt, C. (2004). Dietary boron as a factor in glucose and insulin metabolism. *J Trace Ele* in Experimental Medicinew 17, 258-259.

Iqbal, M.A. (2019). Nano-Fertilizers for Sustainable Crop Production under Changing Climate: A Global Perspective. In M. Hasanuzzaman, M. C. M. T. Filho, M. Fujita, & T. A. R. Nogueira (Eds.), *Sustainable Crop Production*. IntechOpen. https://doi.org/10.5772/intechopen.89089

Jackson, R.D., Idso, S.B., Reginato, R.J. and Pinter, P.J. (1981). Canopy temperature as a crop water stress indicator. *Water Resources Research*, 17(4): 1133–1138.

Jackson, R.D. (1986). Remote sensing of biotic and abiotic plant stress. *Annual Review of Phytopathology*, 24: 265–286.

Jadhav, P.V., Magar, S., Thakur, P., Moharil, M.P., Yadav, H. and Mandlik, R. (2020). Biofortified fodder crops: An approach to eradicate hidden hunger. In *Advances in Agri-Food Biotechnology;* Springer Nature: Singapore, 2020.

Ji, L. and Peters, A.J. (2003). Assessing vegetation response to drought in the northern Great Plains using vegetation and drought indices. *Remote Sensing of Environment*, 87(1): 85–98.

Johnson, G. A., Cardina, J. and Mortensen, D.A. (1997). Site–specific weed management: Current and future direction. In The State of Site-Specific Management for Agriculture, pp. 131–147.

Justice, C. O., Townshend, J. R. G., Vermata, E. F., Masuoka, E., Wolfe, R. E., Saleons, N., Ray, D. P. and Morisette, J. T. (2002). An overview of MODIS Land data processing and product status. *Remote Sensing of Environment*, 83(1-2): 3 – 15.

Katyal, J.C. and Agarwala, S.C. (1982). Micronutrient research in India. *Fert News* 27(2): 67-86.

Kaur, J., Bhatti, D.S. and Goyal, M. (2015). Influence of copper application on forage yield and quality of oats fodder in copper deficient soils. *Indian Journal of Animal Nutrition*, 32(3): 290–294.

Kingra, P. K., Majumder, D. and Singh, S.P. (2016). Application of Remote Sensing and GIS in Agriculture and Natural Resource Management Under Changing Climatic Conditions. *Agricultural Research Journal*, 53(3):295-302.

Kiss, S. and Simihaian, M. (2002). Improving efficiency of urea fertilizers by inhibition of soil urease activity. Kluwer Academic Publishers, Dordrecht, The Netherlands.

Klikocka, H. and Marks, M. (2018). Sulphur and nitrogen fertilization as a potential means of agronomic biofortification to improve the content and uptake of microelements in spring wheat grain DM. *Journal of Chemistry*, 2018(2):1-12.

Kogan, F.N. (1995). Application of vegetation index and brightness temperature for drought detection. *Advances in Space Research*, 15(11): 91–100.

Koshino, M. (1993). The environmental protection framework concerning fertilizer use in Japan. Publisher: National Institute of Agro-Environmental Sciences, Department of Farm Chemicals, Tsukuba, Japan.

Kowalska, G., Kowalski, R., Hawlena, J. and Rowi´nski, R. (2020). Seeds of oilseed rape as an alternative source of protein and mineral. J. Elementol, 25, 513–522.

Kumar, D., Dhaliwal, S.S., Uppal, R.S. and Ram, H. (2016). Influence of nitrogen, zinc and iron fertilizer on growth parameters and yield of Parmal rice in transplanted condition. *Indian Journal of Ecology*, 43(1):115–118.

Kumar, D., Dhaliwal, S.S., Naresh, R.K. and Salaria, A. (2018). Agronomic biofortification of paddy through nitrogen, zinc and iron fertilization: A review. International Journal of Current Microbiology and Applied Sciences, 7(07):2942–2953.

Ladha, J.K., Dawe, D., Pathak, H., Padre, A.T., Yadav, R.L., Bijay, S., Singh, Y., Singh, P., Kundu, A.L., Sakal, R., Ram, N., Regmi, A.P., Gami, S.K., Bhandari, A.L., Amin, R., Yadav, C.R., Bhattarai, E.M., Das, S., Aggarwal, H.P., Gupta, R.K. and Hobbs, P.R. (2003). How extensive are yield declines in long-term rice–wheat experiments in Asia? *Field Crops Research*, 81(2-3):159-180.

Lamb, D. W., M. M. Weedon, and L. J. Rew. (1999). Evaluating the accuracy of mapping weeds in seeding crops using airborne digital imaging: *Avena* spp. in seeding triticale. *Weed Research*, 39(6): 481–492.

Larson, BA. and Frisvold, GB. (1996). Fertilizers to Support Agricultural Development in Sub-Saharan Africa: What is needed and Why? Food Policy. 21(6):509-525.

Lean, M.E. (2019). Principles of human nutrition. *Medicine*, 47(3):140–144.

Lee, W., Alchanatis, V., Yang, C., Hirafuji, M., Moshou, D. and Li, C. (2010). Sensing technologies for precision specialty crop production. *Computer and Electronic in Agriculture*, 74(1): 2-33.

Liaghat, S. and Balasundram, S. K. (2010). A Review: The Role of Remote Sensing in Precision Agriculture. *American Journal of Agricultural and Biological Sciences*, 5(1): 50-55.

Liscano J. F., Wilson C. E., Norman R. J. and Slaton N. A. (2000). Zinc Availability to Rice from Seven Granular Fertilizers. *AAES Res Bulletin*, 963, 1–31.

Liu, Y., Tong, Z. and Prud'homme, R.K.(2008). Stabilized polymeric nanoparticles for controlled and efficient release of bifenthrin. *Pest Management Science*.64:808-812.

López-Alonso, M. (2012). Trace minerals and livestock: Not too much not too little. ISRN *Veterinary Science*, 2012: 704825.

Ma, L., Yang, L., Zhang, M., Yang, Y.Ch. and Chen, B.Ch. (2007). Volatilization of nitrogen of coating controlled-release fertilizer and common fertilizer. (Chinese) *Acta Pedologica Sinica*.

Malewar, G.U. and Ismail, S. (1995).Iron research and agricultural production. In: Tandon HLS, editor. Micronutrient research and agricultural production. New Delhi: India Fertiliser Development and Consultation Organisation; pp. 57–82.

Malveaux, C., Hall, S.G. and Price, R. (2014). Using drones in agriculture: unmanned aerial systems for agricultural remote sensing applications. 2014 Montreal. American Society of Agricultural and Biological Engineers, Quebec Canada, p. 1. July 13–July 16, 2014.

Manzeke, M.G., Mtambanengwe, F., Nezomba, H., Watts, M.J., Broadley, M.R. and Mapfumo, P. (2017). Zinc fertilization increases productivity and grain nutritional quality of cowpea (*Vigna unguiculata* [L.]Walp.) under integrated soil fertility management. *Field Crops Research*, 213:231–244.

McCarrison, R. (1921). Studies in Deficiency Disease. Hazell, Watson and Viney Ltd., London, UK.

Menon, A.R.R. (2012). Remote sensing applications in agriculture and forestry. A paper from the proceedings of the Kerala environment congress, pp. 222-235.

Mikkelsen, R.L., Williams, H.M. and Behel, A.D., Jr. (1994). Nitrogen leaching and plant uptake from controlled-release fertilizers.

Department Soil Science, State University, Raleigh, NC 27695, USA. Fertilizer Research 37(1), 43-50.

Milani, N., McLaughlin, M.J., Stacey, S.P., Kirby, J.K., Hettiarachchi, G.M. and Beak, D.G. (2012). Dissolution kinetics of macronutrient fertilizers coated with manufactured zinc oxide nanoparticles. *Jour. Agric. Food Chem.* 60(16):3991–3998.

Molina, M., Aburto, F., Calderon, R., Cazanga, M. and Escudey, M. (2009). Trace element composition of selected fertilizers used in Chile: Phosphorus fertilizers as a source of long-term soil contamination. *Soil Sed Cont* 18:497-511.

Moore, W.P. (1993). Reacted Layer Technology for Controlled Release Fertilizers. Proceedings: Dahlia Greidinger Memorial International Workshop on Controlled/ Slow Release Fertilizers, Technion – Israel Institute of Technology, Haifa, 7-12 March 1993.

Moran, M.S., Clarke, T.R., Inoue, Y. and Vidal, A. (1994). Estimating crop water deficit using the relation between surface-air temperature and spectral vegetation index. *Remote Sensing of Environment*, 49(3): 246–263.

Moran, M. S., Inoue, Y. and Barnes, E. M. (1997). Opportunities and limitations for image–based remote sensing in precision crop management. *Remote Sensing of Environment*, 61(3): 319–246.

Muralidharudy, Y., Sammi R.K., Mandal, B.N., Subba R.A., Singh, K.N. and Sonekar, S. (2011). GIS-based soil fertility maps of different states of india. AICRP on STCR, Indian Society of Soil Science, Bhopal, pp. 224.

Nayyar. V.K., Takkar, P.N., Bansal, R.L., Singh, S.P., Kaur, N.P. and Sadana, U.S. (1990). Micronutrients in Soils and Crops of Punjab. *Res Bull.* Department of Soils, PAU, Ludhiana.

Nellis, M.D., Pricey, K.P. and Rundquist, D. (2009). Remote Sensing of Cropland Agriculture. The SAGE Handbook of Remote Sensing, (26).

Nicholson, S.E. and Farrar, T.J. (1994). The influence of soil type on the relationships between NDVI, rainfall, and soil moisture in semi arid Botswana: I. NDVI response to rainfall. *Remote Sensing of Environment,* 50(2): 121-133.

Nielsen, F. H. (1990). New essential trace elements for the life sciences. Biol. Trace Elem. Res. 26-27:59941.

Nielsen, F. H. (1992): Nutritional requirements for boron, silicon, vanadium, nickel, and arsenic: current knowledge and speculation. FASEB J. 52661-2667.

Norvell, W. A. (1991). Reactions of metal chelates in soils and nutrient solutions. In: Micronutrients in Agriculture. pp. 187-227.2nd ed. Mortvedt, J. J., Cox, F. R., Shuman, L. M., and Welch, R. M., Eds., Soil Science Society of America, Madison, WI (1991).

Nuss, E.T. and Tanumihardjo, S.A. (2010). Maize: A paramount staple crop in the context of global nutrition. *Food Science and Food Safety,* 9(4): 417–436.

Nyborg, M., Solberg E.D. and Pauly D.G. (1995). Coating of Phosphorus Fertilizers with Polymers Increases Crop Yield and Fertilizer Efficiency. *Better Crops,* 79 (3).

Padbhushan, R., Sharma, S., Kumar, U., Rana, D.S., Kohli, A., Kaviraj, M., Parmar, B., Kumar, R., Annapurna, K., Sinha, A.K. and Gupta, V.V.S.R. (2021). Meta-analysis approach to measure the effect of integrated nutrient management on crop performance, microbial activity, and carbon stocks in indian soils. *Frontiers in Environmental Science,* 9, 724702.

Pepper, I.L., Gerba, C.P., Newby, D.T. and Rice, C.W. (2009). Soil: a public health threat or savior? Critical Reviews in *Env Sc Tech* 39(5): 416-432.

Prasad, A.S., Halstead, J.A. and Nandini, M. (1961). Syndrome of iron deficiency anaemia, hepatosplenolomegaly, dwarfism and geophagia. *American Journal of Medicine,* 31: 532-546.

Prasad, A.S., Miale, A., Farid, Z. and Sandstead, H. (1963). Biochemical studies on dwarfism hypogonadism and anemia. *Arch Int Med* 111(4): 407-428.

Prasad, A. K., Chai, L., Singh, R. P. and Kafatos, M. (2006). Crop yield estimation model for Iowa using remote sensing and surface parameters. *International Journal of Applied Earth Observation and Geoinformation,* 8(1): 26-33.

Prasad, T.N.V.K.V., Sudhakar, P., Sreenivasulu, Y., Latha, P., Munaswamya, V. and RajaReddy, K. (2012). Effect of nanoscale zinc oxide particles on the germination, growth and yield of peanut. *Journal of Plant Nutrition,* 35:905-927

Praveen, K.V. and Aditya, K.S. (2017). Embracing Specialty Fertilisers. *International Journal of Current Microbiology and Applied Sciences,* 6(12): 3865-3868

Pursell Inc. (1992). Coating Thickness Effects. Nutrient Release and Nutrient Analysis.Pursell Industries, Inc., Sylacauga, Alabama, USA.

Pursell Inc. (1994). Temperature Effect on Release of Polyon PCU. Pursell Industries, Inc., Sylacauga, Alabama, USA.

Pursell Inc. (1995). Polyon Polymer Coatings and the RLCTM Process. Pursell Industries, Inc., Sylacauga, Alabama, USA.

Ram, H., Sohu, V.S., Cakmak, I., Singh, K., Buttar, G.S., Sodhi, G.P.S., Gill, H.S., Bhagat, I., Singh, P., Dhaliwal, S.S. and Mavi, G.S. (2015). Agronomic fortification of rice and wheat grains with zinc for nutritional security. Current Science, 109 (6): 1171–1176.

Randhawa, C.S. (1999). Ph.D. Thesis, PAU, Ludhiana.

Rasheed, N., Maqsood, M.A., Aziz, T. and Jabbar, A. (2020). Characterizing lentil germplasm for zinc biofortification and high grain output. *Journal of Soil Sciences and Plant Nutrition*, 20:1336–1349.

Rattan, R.K., Saharan, N. and Datta, S.P. (1999). Micronutrient depletion in Indian soils - extent, causes and remedies. *Fertiliser News*, 44(2), 35-50.

Rebello, C.J., Greenway, F.L. and Finley, J.W. (2014). Whole grains and pulses: A comparison of the nutritional and health benefits. *Journal of Agricultural and Food Chemistry*, 62(29): 7029–7049.

Rego., T.J., Wani, S.P., Sahrawat, K.L. and Pardhasaradhi, G. (2005). Macro-benefits from boron, zinc and sulfur application in Indian SAT: A step for Grey to Green Revolution in agriculture. Global Theme on Agroecosystems Report no. 16. Patancheru 502 324, Andhra Pradesh, India: International Crops Research Institute for the Semi-Arid Tropics. 24 pp.

Rehman, A. and Farooq, M. (2013). Boron application through seed coating improves the water relations, panicle fertility, kernel yield, and biofortification of fine grain aromatic rice. *Acta Physiologiae Plantarum*, 35(2):411–418.

Rehman, A. and Farooq, M. (2016). Zinc seed coating improves the growth, grain yield and grain biofortification of bread wheat. *Acta Physiologiae Plantarum*. 38(10): 238.

Rehman, M., Muhammad, F., Muhammad, N., Ahmad, N. and Babar, S. (2018). Seed priming of Zn with endophytic bacteria improves the productivity and grain biofortification of bread wheat. *European Journal of Agronomy,* 94: 98–107.

Remedios, C., Ros'ario, F. and Bastos, V. (2012). Environmental Nanoparticles Interactions with Plants: Morphological, Physiological, and Genotoxic Aspects. *Journal of Botany,* Volume 1-9.

Rico, C.M., Majumdar, S., Duarte-Gardea, M., Peralta-Videa, J.R. and Gardea-Torresdey, J.L. (2011). Interaction of nanoparticles with edible plants and their possible implications in the food chain. *Jour. Agric. Food Chem.*59(8):3485–3498.

Riedell, W. E., Osborne, S. L. and Hesler, L. S. (2004). Insect pest and disease detection using remote sensing techniques. Proceedings of 7th International Conference on Precision Agriculture and Other Precision Resources Management, Hyatt Regency, Minneapolis, MN, USA, pp.1380-1387.

Rietze, E. and Seidel, W. (1994). An adequate supply of coated slow-acting fertilizers reduces nitrate leaching). Urania Agrochem Germany. Gartenbau Magazin 3 (7), 32-33.

Rook, G.A.W. (2010). 99th Dahlem conference on infection, inflammation and chronic inflammatory disorders: Darwinian medicine and the "hygiene" or "old friends" hypothesis. *Clinical and Experimental Immunology,* 160(1): 70–79.

Rouse, J.W., Haas, R.H., Schell, J.A. and Deering, D.W. (1974). Monitoring Vegetation Systems in the Great Plains with ERTS. Third ERTS-1 Symposium NASA, NASA SP-351, Washington DC, 309-317.

Rubio-Covarrubias, O.A., Brown, P.H., Weinbaum, S.A., Johnson, R.S. and Cabrera, R.I. (2009). Evaluating foliar nitrogen compounds as indicators of nitrogen status in Prunus persica trees. *Scientia Horticulturae,* 120(1):27-33.

Sadana, U.S., Manchanda, J.S., Khurana, M.P.S., Dhaliwal, S.S. and Singh, H. (2010). The Current Scenario and Efficient Management of Zinc, Iron, and Manganese Deficiencies *Better Crops South Asia* 5: 24-26.

Sakal, R., Singh, A.P., Sinha, R.B. and Bhogal, N.S.(1996). Twentyfive years of research on micro and secondary nutrients in soils of Bihar, Rajendra Agricultural University, Pusa Bihar.pp. 208.

Sakamoto, T., Yokozawa, M., Toritani, H., Shibayama, M., Ishitsuka, N. and Ohno, H. (2005). A crop phenology detection method using time series MODIS data. *Remote Sensing of Environment,* 96(3-4): 366–374.

Sandhu, A., Dhaliwal, S.S., Shukla, A.K., Sharma, V. and Singh, R. (2020). Fodder quality improvement and enrichment of oats with Cu through biofortification: A technique to reduce animal malnutrition. *Journal of Plant Nutrition,* 43(3):1-12.

Sarkar, A.K. and Singh, S. (2003). Crop response of secondary and micronutrients in acidic soils of India. *Fertilser News,* 48(4): 47-54.

Seiler, R.A., Kogan, F. and Wei, G. (2000). Monitoring weather impact and crop yield from NOAA AVHRR data in Argentina. *Advances in Space Research,* 26(7): 1177–1185.

Selim, M.M. (2020). Introduction to the integrated nutrient management strategies and their contribution to yield and soil properties. *International Journal of Agronomy,* 2020, Article ID 2821678.

Shanmugapriya, P., Selvaraj, R., Ramesh, T. and Ponnusamy, J.(2019). Applications of Remote Sensing in Agriculture - A Review, *International Journal of Current Microbiology and Applied Sciences,* 8(1): 2270-2283.

Sharma, U and Kumar, P (2016). Micronutrient Research in India: Extent of deficiency, crop responses and future challenges, *International Journal of Advanced Research,* 4:4, 1402-1406.

Shaviv, A. (1996). Plant response and environmental aspects as affected by rate and pattern of nitrogen release from controlled N fertilizers. In: Progress in Nitrogen Cycling Studies. O. Van Cleemput et al. (eds.). 285-291.

Shaviv, A. (2005). Controlled Release Fertilizers. IFA International Workshop on Enhanced-Efficiency Fertilizers, Frankfurt. International Fertilizer Industry Association Paris, France.

Shoji, S. and Gandeza, A.T. (1992): Controlled release fertilizers with polyolefin resin coating. Kanno Printing Co. Ltd. Sendai, Japan.

Shoji, S. (ed) (1999). Meister controlled release fertilizer – Properties and utilisation. Konno Printing Company Ltd. Sendai. Japan.

Shoji, S. (2005). Innovative use of controlled availability fertilizers with high performance for intensive agriculture and environmental conservation. Science in China Ser. C. Life Sciences 48, 912-920.

Shoji, S. and Takahashi, C. (1999). Innovative fertilization methods. In: Meister controlled release fertilizer – Properties and utilisation. Shoji, S. (ed). Konno Printing Company Ltd. Sendai, Japan

Shukla, A.K., Dwivedi, B.S. and Singh, V.K. (2009). Macrorole of micronutrients. *Indian Journal of Fertilisers,* 5(5):11-30.

Shukla, A.K. and Behera, S.K. (2012). Micronutrient fertilizers for higher productivity. *Indian Journal of Fertilisers,* 8(4):100-117.

Shukla, A.K., Behera, S.K., Satyanarayana, T. and Majumdar, K. (2014). Importance of micronutrients in Indian Agriculture. *Indian Journal of Fertilisers,* 10(12): 94-112.

Shukla, A.K., Tiwari, P.K. and Prakash, C. (2014). Micronutrients deficiencies vis-à-vis food and nutritional security of India. *Indian Journal of Fertilisers,*10 (12), 94-112.

Shukla, A.K., Behera, S.K., Lenka, N. and Chaudhary, S.K. (2016). Spatial variability of soil micronutrients in the intensively cultivated Trans-Gangetic Plains of India, *Soil & Till. Res.* 163: 282-289.

Shukla, A.K. and P.K. Tiwari.(2016). Micro and secondary nutrients and pollutant elements research in India. Progress Report 2014-16. AICRP-MSPE, ICAR-IISS, Bhopal. pp.1-196.

Shukla, A.K., Behera, S.K., Lenka, N.K., Tiwari, P.K., Prakash, C., Malik, R.S., Sinha, N.K., Singh, V.K., Patra, A.K. and Chaudhary, S.K. (2016). Spatial variability of soil micronutrients in the intensively cultivated Trans-Gangetic plains of India. *Soil and Till Res*, 163: 282-289.

Shukla, A.K. and Tiwari, P.K. (2016). Micro and Secondary Nutrients and Pollutant Elements Research in India:. Coordinator's Report. *AICRP on Micro- and Sec Nut Poll Ele in Soils and Plants*, ICAR-IISS, Bhopal.

Shukla, A.K. and Behera, S.K. (2017). Micronutrients research in India: Retrospect and prospects. *Preprint, FAI Annual Seminar.* pp. SII-4/1-SII-4/ 17. The Fertiliser Association of India, New Delhi.

Shukla, A.K., Behera, S.K., Pakhre, A. and Chaudhari, S.K.(2018). Micronutrients in Soils, Plants, Animals and Humans, *Indian Journal of Fertizers*, 14 (4):30-54

Shukla, A.K., Behera, S.K. and Majumdar, K.(2019). Importance of micronutrients in Indian Agriculture, Better Crops- South Asia, pp 6- 10.

Sillanpää, M. (1990). Micronutrient Assessment at the Country Level: An International Study. FAO *Soils Bull,* 63. FAO, Rome, Italy.

Silva, M. S., Cocenza, D. S., Grillo, R., Melo, N. F. S., Tonello, P. S., Oliveira, L. C., Cassimiro, D. L., Rosa, A. H. and Fraceto, L.F. (2011). Paraquat-loaded alginate/chitosan nanoparticles: Preparation, characterization and soil sorption studies. *Journal of Hazardous Materials*,190(1-3): 366-374.

Silveira, M. L., Obour, A. K., Vendramini, J. M. B. and Sollenberger, L. E. (2011). Using tissue analysis as a tool to predict bahiagrass phosphorus fertilization requirement. *Journal of Plant Nutrition*, 34: 2193–2205.

Singh, B. and Randhawa, S.S. (1990). In National Symposium. 9th Annual Convention of Indian Society of Veterinary Medicines, Hyderabad.

Singh, M.V. (2001). Evaluation of current micronutrient stocks in different agroecological zones of India. *Fertiliser News*, 46(2), 25-42.

Singh, A.P.; Singh, M.V.; Sakal, R.; Chaudhary, V.C. (2006). Boron nutrition of crops and soils of Bihar. *Tech. Bull.*, 6, 1–80.

Singh, M. V., (2008). Micronutrient Deficiencies in Crops and Soils in India In: Micronutrient deficiencies in Global Crop Production (Ed.) B. J. Alloway.

Singh, M.V. (2009). Micronutrient nutritional problems in soils of India and improvement for human and animal health. *Indian Journal of Fertilisers*, 5(4), 11-26.

Singh, P., Dhaliwal, S.S. and Sadana, U.S. (2013). Iron enrichment of paddy grains through ferti-fortification. *Journal of Research Punjab Agricultural University*, 50(1-2):32–38.

Singh, N.B., Amist, N., Yadav, K., Singh, D., Pandey, J.K. and Singh, S.C. (2013). Zinc oxidenanoparticles as fertilizer for the germination, growth and metabolism of vegetable crops. *Jour Nanoengg Nanomfg.* 3:353-364.

Singh, V., Bhatnagar, A. and Singh, A.P. (2016). Evaluation of leaf-colour chart for need-based nitrogen management in maize (Zea mays) grown under irrigated condition of Mollisols. *Indian Journal of Agronomy*, 61(1):47-52

Singh, V.P., Singh, G. and Dhaliwal, S.S. (2019). Agronomic biofortification of chickpea with zinc and iron through application of zinc and urea. *Communications in Soil Science and Plant Analysis*, 50(15): 1864–1877.

Sinha, B.P., Jha, J.G.J. and Sinha, B.K. (1976). Leucoderma in Indian buffaloes. *Ind Vet J* 53: 812-815

Stein, A.J. and Qaim, M. (2007). The human and economic cost of hidden hunger. *Food and Nutrition Bulletin*, 28(2):125–34.

Stevens, G.A., Finucane, M.M., De-Regil, L.M., Paciorek, C.J., Flaxman, S.R., Branca, F., Peña-Rosas, J.P., Bhutta, Z.A. and Ezzati, M. (2013). Global, regional, and national trends in haemoglobin concentration and prevalence of total and severe anaemia in children and pregnant and non-pregnant women for 1995-2011: a systematic analysis of population-representative data, *Lancet Global Health*, 1(1):e16-25

Stewart, W.M., Dibb, D.W., Johnston, A. E. and Smyth, T.J. (2005). The Contribution of Commercial Fertilizer Nutrients to Food Production, *Agr.Jour*, 97(1):1-6.

Subramanian, K.S., Manikandan, A., Thirunavukkarasu, M. and Sharmila, R.C.(2015). Nano-fertilizers for balanced crop nutrition. In: Rai M, Ribeiro C, Mattoso L, Duran N, editors. Nanotechnologies in Food and Agriculture. Switzerland: Springer International Publishing; pp. 69-80.

Száková, J., Praus, L., Tremlová, J., Kulhánek, M. and Tlustoš, P. (2017). Efficiency of foliar selenium application on oilseed rape (*Brassica napus* L.) as influenced by rainfall and soil characteristics. *Archives of Agronomy and Soil Science*, 63(9):1240–1254.

Takkar, P.N. and Nayyar, V.K. (1984). Response of Crops to Micronutrient Applications. Proc. FAI (NRC). Seminar, Jaipur. 95-123.

Takkar, P.N., Chhibba, I.M., and Mehta S.K (1989). Twenty Years of Coordinated Research on Micronutrients in Soils and Plants, Indian Institute of Soil Science, Bhopal IISS, Bulletin 1. pp 394.

Takkar, P.N. (1996). Micronutrient research and sustainable agricultural productivity in India. *Journal of the Indian Society of Soil Science,* 44(4): 562–581

Takkar, P.N. (2011). A roundtable meeting on boron in Indian agriculture: Keynote address. *Indian Journal of Fertilisers,* 7(4):142-148.

Takkar, P.N. (2015). Long-term effect of rice-wheat cropping system on cobalt in relation to manganese and iron content in coarse textured calcareous alluvial soils Proceedings of Indian National Science Academy, 81(3): 663-682.

Takkar, P.N. and Shukla, A.K. (2015). In State of Indian Agriculture – Soil. H.Pathak, S.K. Sanyal and P.N. Takkar (eds.). NAAS, New Delhi, India, pp. 121-152.

Tandon HLS. (2009). Micronutrient Hand book- from research to practical application. Fert. Dev. and Consultation Org., New Delhi, India, 19-27.

Tandon, H.L.S. (2012). Fertilizer management: Balance-efficiency-profitability. Fertilizer development and consultation organization, New Delhi. pp. 187.

Tarafdar, J.C., Sharma, S. and Raliya, R. (2013). Nanotechnology: Interdisciplinary science of applications. *African Jour Biotech.* 12(3): 219-226.

Taylor, H. M., Upchurch, D. R. and McMichael, B.L (1990). Applications and limitations of rhizotrons and minirhizotrons for root studies. *Plant Soil,*129:29-35 (1990)

Thenkabail, P.S., Smith, R.B. and De-Pauw, E. (2002). Evaluation of narrowband and broadband vegetation indices for determining optimal hyperspectral wavebands for agricultural crop characterization. *Photogrammetric Engineering and Remote Sensing*, 68(6): 607–621.

Thenkabail, P. S., Enclona, E. A., Ashton, M. S. and Van Der Meer, B. (2004). Accuracy assessments of hyperspectral waveband performance for vegetation analysis applications. *Remote Sensing of Environment*, 91 (3–4): 354–376.

Thilly, C., Lagasser, R. and Roger, G (1980). Impaired fetal and postnatal development and high perinatal death rate in a severe iodine deficient area. In: Stockigt JR et al., eds. Thyroid Research VIII. Proceedings of the 8th International Thyroid Congress, Canberra, Australian Academy of Research, 20–23.

Thompson, B. and Amoroso, L. (2011). Combating Micronutrient Deficiencies: Food-Based Approaches, *Food and Agriculture Organization of the United Nations; CABI.*

Tiwari, K.N., Sharma, S.K., Singh, V.K, Dwivedi, B.S. and Shukla, A.K., (2006). Site-specific Nutrient Management for Increasing Crop Productivity in India: Results with Rice-Wheat and Rice-Rice System, pp. 92, PDFSR, Modipuram and PPIC-India Programme, Gurgaon.

Tiwari, K.N. (2007). Breaking yield barriers and stagnation through site-specific nutrient management. Journal of Indian Society of Soil Science, 55(4):444-454.

Tosukhowong, P., Tungsanga, K., Eiam-Ong, S. and Sitprija, V. (1999). Environmental distal renal tubular acidosis in Thailand: an enigma. American Journal of Kidney Diseases 33(6): 1180-1186.

Ullrich-Eberius, C.I., Sanz, A. and Novacky, A. J. (1989). Evaluation of arsenate and vanadate associated changes of electrical membrane potential and phosphate transport in Lemma Gibba G1. *J Exp Bot* 40(1): 119-128.

Van Campen, D. R. (1991). Trace elements in human nutrition. In: Micronutrients in Agriculture. pp. 663-701. 2nd ed. Mortvedt, J. J., Cox, F. R., Shuman, L. M., and Welch, R. M., Eds., Soil Science Society of America, Madison, WI.

Vasudevan, V. (1987). In Proceedings of Symposium on Recent Advances in Mineral Nutrition (V.M. Mandokhot, Ed.), CCS HAU, Hisar.

Venkidasamy, B., Selvaraj, D., Nile, A.S., Ramalingam, S., Kai, G. and Nile, S.H. (2019). Indian pulses: A review on nutritional, functional and biochemical properties with future perspectives. *Trends in Food Science and Technology*, 88: 228–242.

Verma, P., Chauhan, A. and Ladon, T.(2020). Site specific nutrient management: A review. *Research Journal of Pharmacognosy and Phytochemistry*, 9(5): 233-236.

Vidal, A., Pinglo, F., Durand, H., Devaux-Ros, C. and Maillet, A. (1994). Evaluation of a temporal fire risk index in Mediterranean forests from NOAA thermal IR. *Remote Sensing of Environment*, 49(3): 296–303.

Waleed, F.A. (2018). Nanotechnology Application in Agriculture, *Acta Scientific Agriculture (ISSN: 2581-365X)* 2(6): 99-102

Walker, R., Morris, S., Brown, P. and Gracie, A. (2004). Evaluation of potential for chitosan to enhance plant defence. Australian Government, RIRDC Report No.4: 49.

Wallace, J.F., Caccetta, P.A. and Kiiveri, H.T. (2004). Recent developments in analysis of spatial and temporal data for landscape qualities and monitoring. *Australian Ecology*, 29(1):100-107.

Wallace, R.J., Loffi, J.M., Vance, S.M., Jacob, J., Dunlap, J.C. and Mitchell, T.A. (2018). Pilot visual detection of small unmanned aircraft systems (sUAS) equipped with strobe lighting. *Journal of Aviation Technology and Engineering,* 7(2): 57-66.

Wang, F. (1996). Modelling Nitrogen Transport and Transformations in Soils Subject to Environmentally Friendly Fertilization Practices. Research thesis. Technion – Israel Institute of Technology, Haifa, Israel.

Wang, J., Price, K.P. and Rich, P.M. (2001). Spatial patterns of NDVI in response to precipitation and temperature in the central Great Plains. *International Journal of Remote Sensing,* 22(18): 3827–3844.

Welch, R.M., Combs, G.F. Jr. and Duxbury, J.M. (1997). Toward a greener revolution. *Issues in Science and Technolology* 14, 50-58.

Wessells, K.R. and Brown, K.H. (2012). Estimating the global prevalence of zinc deficiency: results based on zinc availability in national food supplies and the prevalence of stunting. PLoS One,7(11):e50568.

Wilhite, D. A. and Glantz, M. H. (1985). Understanding the drought phenomenon: The role of definitions. *Water International,* 10(3): 111-120.

Xiao, Q., Zhang, F.D. and Wang, Y.J. (2008). Effects of slow/controlled release fertilizers felted and coated by nano-materials on nitrogen recovery and loss of crops. *Plant Nutrition and Fertilizer Science,* 14:778-784.

Yadav, R.L., Dwivedi, B.S., Prasad, K., Tomar, O.K., Shurpali, N.J. and Pandey P.S. (2000): Yield trends and changes in soil organic-C and available NPK in a long-term rice-wheat system under integrated use of manures and fertilisers. *Field Crops Research,* 68(3): 219–246.

Yoo, H.S., Kim, T.G. and Park, T.G.(2009). Surface-functionalized electrospun nanofibers for tissue engineering and drug delivery. *Advanced Drug Delivery Reviews,* 61:1033-1042.

Zafar, S., Li, Y.L., Li, N.N., Zhu, K.M. and Tan, X.L. (2019). Recent advances in enhancement of oil content in oilseed crops. *Journal of Biotechnology*, 301: 35–44.

Zambryski, P. (2004). Cell-to-cell transport of proteins and fluorescent tracers via plasmodesmata during plant development. *J Cell Biol.* 164(2): 165–168.

Zhang, M. (2007). Effect of coated controlled-release fertilizer on yield increase and environmental significance. (Chinese) Ecology and Environment.

Zhang, Q.L., Zhang, M. and Tian, W.B. (2001). Leaching characteristics of controlled release and common fertilizers and their effects on soil and ground water quality. (Chinese) Soil and Environmental Sciences, 10(2), 98-103.

Zhang, C. and Kovacs, J.M. (2012). The application of small unmanned aerial systems for precision agriculture: a review. Precision Agriculture 13(6):693–712.

Ziegler, E.E. and Filer, L.J. (1996). Present Knowledge in Nutrition. 7th Edition. International Life Sciences Institute – Nutrition Foundation, Washington DC, USA.

TRIAL REPORT 1

Name of the Organisation : National Horticultural Research and Development Foundation, Nasik

Name of the Implementing Centre : Regional Research Station, Chitagaon, Nashik(Maharashtra)
: Regional Research Station, Karnal, Haryana

Name of Company : Aries Agro Limited

Product : EDTA chelated Manganese in liquid form

Crop and Season : Cucumber, Summer 2017

Details of Experiment

Experimental Design : Randomised Block design (RBD)
Number of treatments : Seven
Treatments

Sr. No	Treatments	Dose
1	Application of chelated manganese liquid through drip/ drenching on 20th DAS	500 ml/acre
2	Application of chelated manganese liquid through drip/ drenching on 45th DAS	500 ml/acre
3	Foliar application of chelated manganese liquid through drip/ drenching on 20th DAS	2 ml/litre of water
4	Foliar application of chelated manganese liquid through drip/ drenching on 45th DAS	2 ml/litre of water
5	Application of chelated manganese liquid through drip/ drenching and foliar application on 20th DAS	500 ml/acre through drip and 2 ml/litre of water for spray

6	Application of chelated manganese liquid through drip/ drenching and foliar application on 45th DAS	500 ml/acre through drip and 2 ml/litre of water for spray
7	Application of chelated manganese liquid through drip/ drenching and foliar application on 20th and 45th DAS	500 ml/acre through drip and 2 ml/litre of water for spray
8	Control	----

Number of Replications	: Three
Bed Size	: 7.0 X 1.5 m (Chitegaon); 3.6 X 1.2 m (Karnal)
Spacing	: 1.5 X 0.5 m (Chitegaon); 90 X 60 cm (Karnal)

Observations

Total Pickings	:10 at Chitegaon and 13 at Karnal
Parameters analysed	: Height of plant, leaf size, days for flower initiation, days for first harvest, diameter and length of fruits, weight of 10 fruits, number of fruits per plant, number of pickings and shelf life at ambient condition and yield

Results:

Chitegaon (Nashik)

The data presented in table no 1 revealed that all the recorded characters except weight of 10 fruits, number of fruits per plant and yield per hectare showed non significant difference between the treatments. Significantly highest weight of 10 fruits (1.06 kg), number of fruits per plant (15.27) and yield per hectare (242.30 q) were recorded in the treatment T7 (Application of Mn - Chel liquid through drip /drenching & foliar application on 20th & 45th DAT) and found at par with the treatments T2 (0.94 kg), T3 (0.93 kg) T4 (0.97 kg), T5 (1.00 kg) and T6 (1.06 kg) in respect of weight of 10 fruits, with treatment T6 (15.07) in respect of number of fruits per plant and with treatments T2 (229.28 g), T3 (227.55 g), T4 (229.42 al. T5 (233.45 q) and T6 (238.98 q) in respect of yield per ha.

Salaru (Karnal)

The data presented in table no 2 revealed that all the recorded characters except yield per hectare showed non-significant differences between given treatments. Significantly highest yield (280.01 q/ha) was recorded in the treatment T3 (Foliar application of Mn-Chel liquid @ 2.0 ml/lit on 20th DAS) and found at par with the treatment 17 (263.31 q/ha). Overall maximum length of fruit (14.84 cm), weight of 10 fruits (1.38 kg) and number of fruits per plant (8.98) were observed in the treatment T3.

Treatments:

T1 - Application of chelated manganese liquid through drip/ drenching on 20th DAS @500 ml/acre

T2 - Application of chelated manganese liquid through drip/ drenching on 45th DAS @500 ml/acre

T3 - Foliar application of chelated manganese liquid through drip/ drenching on 20th DAS @ 2 ml/litre of water

T4 - Foliar application of chelated manganese liquid through drip/ drenching on 45th DAS @ 2 ml/litre of water

T5 - Application of chelated manganese liquid through drip/ drenching and foliar application on 20th DAS @500 ml/acre through drip and 2 ml/litre of water for spray

T6 - Application of chelated manganese liquid through drip/ drenching and foliar application on 45th DAS @500 ml/acre through drip and 2 ml/litre of water for spray

T7 - Application of chelated manganese liquid through drip/ drenching and foliar application on 20th and 45th DAS @500 ml/acre through drip and 2 ml/litre of water for spray

T8 - Control

Table 1. "Efficacy of EDTA chelated Manganese in liquid form on growth and yield of Cucumber (Cucumis sativus)' at RRS Nashik during Rabi, 2016-17

TREATMENT	Plant Height At 60DAS (cm)	Leaf Size (cm²)	Days for Flower Initiation	Days for first harvesting	Length of fruit (cm)	Diameter of fruit (cm)	Weight of 10 fruits (kg)	No of fruit per plant	Number of picking	Yield (q/ha)	Shelf Life (days)
T_1	98.13	75.50	46.67	52.67	12.50	3.77	0.89	11.83	10.00	216.14	4
T_2	98.20	75.63	47.00	54.67	12.83	3.83	0.94	12.37	10.00	229.28	4
T_3	98.07	75.50	47.33	53.33	12.53	3.77	0.93	12.10	9.67	227.55	4
T_4	98.60	77.18	47.33	53.33	12.30	3.67	0.97	12.77	10.00	229.42	4
T_5	98.47	77.70	48.00	54.00	12.60	3.77	1.00	12.80	9.67	233.45	4
T_6	100.13	77.93	46.67	52.67	13.53	3.90	1.05	15.07	10.00	238.98	4
T_7	101.67	78.50	46.00	52.33	13.90	3.97	1.06	15.27	10.00	242.30	5
T_8	97.47	71.46	47.33	53.33	12.17	3.67	0.84	10.73	9.67	203.53	4
S.E.m±	1.61	4.06	1.15	1.06	0.81	0.25	0.06	0.95	0.31	10.27	0.30
C.D at 5%	NS	NS	NS	NS	NS	NS	0.13	2.04	NS	22.03	NS

Table 2: "Efficacy of EDTA chelated Manganese in liquid form on growth and yield of Cucumber (Cucumis sativus)'at RRS Karnal during Rabi, 2016-17

TREATMENT	Plant Height At 60DAS (cm)	Length of fruit (cm)	Diameter of fruit (cm)	Weight of 10 fruits (kg)	No of fruit per plant	Number of picking	Yield (q/ha)	Shelf Life (days)	Fruit Colour	Micro Nutrient deficiency
T_1	148.67	14.68	3.56	1.28	8.90	12.33	237.53	5	Light Green	Nil
T_2	148.00	14.80	3.55	1.33	7.73	12.00	235.84	5	Light Green	Nil
T_3	142.00	14.84	3.61	1.38	8.98	12.00	280.01	5	Light Green	Nil
T_4	139.33	14.72	3.59	1.25	7.85	12.33	219.83	5	Light Green	Nil
T_5	137.67	14.70	3.51	1.27	7.72	12.33	203.51	5	Light Green	Nil
T_6	131.00	14.73	3.57	1.25	8.12	12.67	214.66	5	Light Green	Nil
T_7	136.33	14.77	3.78	1.29	8.83	12.67	263.31	5	Light Green	Nil
T_8	129.00	14.42	3.45	1.16	7.71	12.00	188.54	5	Light Green	Nil
S.E.m+	10.04	0.43	0.14	0.03	0.59	0.55	11.58	-	-	-
C.D at 5%	NS	NS	NS	Ns	Ns	NS	24.84	-	-	-

Conclusion:

The study conducted at RRS, Chitegaon, Nashik on Cucumber variety Shiny during kharif season revealed that soil application of chelated Manganese in liquid form through drip or drenching as well as foliar application at 20th and 45th DAS enhanced yield significantly over control in Cucumber at Nadhik (MS) condition.

The study conducted at RRS, Karnal, on Cucumber variety Malav (F1 hybrid) during kharif season revealed that foliar application of chelated Manganese in liquid form at 20th DAS increases yield significantly over control in Cucumber under Karnal (Haryana) conditions.

TRIAL REPORT 2

Name of the Organisation : Mahatma Phule Krishi Vidyapeeth, Rahuri

Name of the Implementing Centre : AICRP on Vegetable Crops, Department of Horticulture, MPKV., Rahuri.

Name of Firm : M/s Agro Inputs Manufacturers Association (of India), Pune, Maharashtra, India

Product : Manganese Chelate (12%)

Crop and Season : Okra, Kharif

Details of Experiment

Experimental design : Randomised Block design
No. of replications : Five
Plot size : 7.2 sq. mtr
Soil type : Medium black
No. of Treatments : Four
Treatment details

Treatment No	Treatments	Dose (g/lit)
1	Manganese Chelate (12%)	0.5
2	Manganese Chelate (12%)	1.0
3	Manganese Chelate (12%)	1.5
4	Control	

Crop Details

Crop/Variety : Phule Utkarsha
Date of sowing : 28.02.2014
Spacing : 60x30cm
Date of harvestings : 16.04.2014 to 26.05. 2014

Methodology

Spraying equipment		: Knapsack sprayer
Method of Application		: Foliar application

Time of Application		: Three sprays
    a. At 21-30 days after sowing
    b. At 50% of flowering
    c. At fruit development stage

Observations taken		: i) Plant ht (cm)
    ii) No. of fruits/plant
    iii) Fruit weight (g)
    iv) Fruit length (cm)
    v) Fruit diameter (cm)
    vi) Yield /plot (kg)
    vii) Yield (q/ha)
    viii) Incidence of diseases and Pest

Results:

The Okra variety "Phule Utkarsha" was used to study the effect of Manganese Chelate (12%) sprays of different concentrations on yield during summer 2014. The observations on growth and yield parameters were given in Table 1. Among the different treatments studied, the maximum fruit yield was observed in treatment T2 (170.18q /ha) i.e. Manganese Chelate (12 %) at 1.0 g per liter at 30 days after sowing, 50% flowering and at fruit development stage which was at par with the treatment T3 (163.95q /ha) i.e. Manganese Chelate (12 %) at 1.5 g per liter at 30 days after sowing, 50 % flowering and at fruit development stage. The lowest fruit yield was observed in T4 (148.55q /ha) i.e. control treatment followed by T1 (150.14q /ha) i.e. Manganese Chelate (12%) at 0.5 g per liter at 30 days after sowing,50 % flowering and at fruit development stage All the doses of Manganese Chelate (12%) recorded the maximum fruit yield in okra than the only control treatment.

Table 1: Effect of different concentrations of Manganese Chelate (12 %) on growth and yield of Okra

Treat No	Plant ht. (m)	No. of fruits/ plant	Fruit length (cm)	Fruit Diameter (cm)	Av. wt. (g)	Fruit colour	Yield/ plant (g)	Kg/ plot	q/ha	YVMV Incidence %	Fruit Borer %
T₁	75.02	23.16	12.12	1.43	12.2	Light green	360.58	10.82	150.44	24.98	33.35
T₂	76.58	27.2	12.75	1.44	13.08	Light green	408.71	12.26	170.18	15.72	32.33
T3	73.23	24.36	12.58	1.44	12.52	Light green	393.72	11.81	163.95	18.46	30.12
T4 (Control)	77.01	22.08	11.4	1.42	11.05	Light green	356.77	10.7	148.55	26.13	35.45
SE ±	NS	0.87	0.29	NS	0.29	--	7.22	0.21	3	--	--
CD at 5%	NS	2.69	0.89	NS	0.89	--	22.26	0.66	9.27		

Conclusion

In study of Manganese Chelate (12 %) sprays at different concentrations in Okra during summer2014, the treatment T_2 i.e. Manganese Chelate(1 2 %) at 1.0 g per liter at 30 days after sowing, 50% flowering, and at fruit development stage recorded the maxirnum fruit yield of (170.18q/ha.)which was at par with the treatment T3 (163.95q/ha) i.e. Manganese Chelate(12%) at 1.5g per liter at 30 days after sowing, 50 % flowering and at fruit development stage.

TRIAL REPORT 3

Name of the Organisation	: Dr. B.S. Konkan Krishi Vidyapeeth, Dapoli
Title of the project	: Effect of Manganese chelate (12%) on growth and yield of Okra
Name of Firm	: M/s Agro Inputs Manufacturers Association (of India), Pune, Maharashtra, India
Treatment Details	: - T1- Spraying of Manganese Chelate @ 0.5 g /L T2 - Spraying of Manganese Chelate @1.0 g/L T3 - Spraying of Manganese Chelate @1.5 g /L T4 - Control (Recommended Package of practices)
Replications	: Five (5)
Design	: Randomised Block Design (RBD)
Plot Size	: 3.0m X 2.7 m
Observations	: 1.Plant Height (cm) 2.No. of branches 3.Length of fruit (cm) 4.Girth of fruit (cm) 5.Average wt. of fruit (kg) 6. No. of fruits per plant 7.No. of fruits per plot 8.Yield per plant (Kg) 9 Yield per plot (kg) 10 Yield per hectare (t)

Results

Table No 1 Effect of Manganese chelate (12%) on growth of Okra

Treatment	Plant Height (Cm)	No.Of Branches	Length Of Fruit (Cm)	Girth Of Fruit(Cm)	Avg.Fruit Wt.(Kg)
T_1	106.50	1.30	16.496	18.546	0.020
T_2	106.78	1.36	16.480	18.174	0.019
T_3	103.42	1.34	17.232	18.366	0.055
T_4	100.53	1.24	16.028	17.234	0.018
S.E+	3.32	0.11	0.44	0.38	0.017
C.D @ 5%	NS	NS	1.37	1.20	0.152

Table No 2 Effect of Manganese chelate (12%) on yield attributing characters of Okra

Treatment	No Of Fruits Per Plant	No Of Fruits Per Plot	Yield Per Plant(G)	Yield Per Plot (Kg)	Yield Per Hectare (T)
T_1	17.04	613.40	0.243	8.75	10.80
T_2	19.36	701.40	0.296	10.65	13.15
T_3	2.39	770.20	0.321	11.41	14.09
T_4	16.71	601.60	0.274	9.88	12.19
S.E+	1.12	41.10	0.014	0.85	0.56
C.D@ 5%	3.46	126.64	0.043	2.63	1.71

Results:

The results indicated that the height and number of branches was not influenced significantly due to spraying of ferrous chelate at various concentrations. The number of fruits per plant affected significantly and maximum number of fruits per plant (21.39) and per plot (770.20) were recorded maximum in treatment T3 - Spraying of Manganese Chelate (a), 1.5 g/L. The yield per plant and yield per plot was affected significantly due to spraying of ferrous chelate at various concentrations. The spraying of Manganese Chelate @ 1.5 g/l (T3) recorded maximum yield per plant (0.321 kg). per plot (11.41 kg) and per hectare (14.09t).

TRIAL REPORT 4

Name of the Organisation	: Vasantrao Naik Marathwada Krishi Vidyapeeth, Parbhani
Name of the Implementing Centre	: Horticulture Research Scheme (Vegetable) Sub Campus, VNMKV, Parbhani-431 402
Product	: Manganese Chelate (12%)
Name of Firm	: Agro Inputs Manufacturers Association (India)
Test Season	: Kharif 2015-2016

Material and Methods:

Variety	: Okra- Parbhani Bhendi (PBN-OK-1)
No. of Treatments	: Four
Treatment Details	: T1- Manganese Chelate 12% @0.5g/L
	T2- Manganese Chelate 12% @1.0g/L,
	T3- Manganese Chelate 12% @1.5g1L
	T4- Control
No. of Replications	: Five
Experimental Design	: Randomized Block Design
Plot Size	: 3.0X2.7 m
Spacing	: 60 X 30 cm
Date of Seeds sowing	: 13.07.2 015
Soil Type	: Medium black cotton soil
Fertilizer dose	: 100:50:50
Application Method	:

First Spray-21-30 Days after sowing/transplanting(peak growth)
Second Spray – 50% Flowering stage
Third Spray,:- Fruit development stage

Result:

1. Effect of different levels of Manganese chelate 12% on growth of okra

It is evident from the table I that different levels of Manganese chelate 12% showed difference on vegetative growth parameters of okra.

Regarding vegetative parameters, maximum plant height (87.9 cm) was recorded in treatment applied with Manganese chelate 12% of 1g per litre while more number of branches per plant (2.7) and number of leaves per plant (65.4) was recorded in treatment of Manganese chelate 12% of 1.5g per litre. The maximum number of fruits per plant (40.5) was recorded in treatrnent applied with Manganese chelate 12% of 1g per litre.

Table 1 Effect of different levels of Manganese chelate 12% on growth of okra

Treatments	Plant height (cm)	No. of Branches	No. of leaves	No. of fruit/ plant	Plant Spread	
					(E.W)	(S.N.)
T1 Manganese Chelate 12% @ 0.5 g/l	87.6	2.4	64.8	38.8	41.1	42.9
T2 Manganese Chelate 12% @ 1.0 g/l	87.9	2.6	64.1	40.5	44.7	46.4
T3 Manganese Chelate 12% @ 1.5 g/l	84.4	2.7	65.4	36.5	44.5	45.5
T4 Control	79.8	2.4	60.7	35.9	45.1	42.4
SE	1.25	0.22	3.84	0.41	1.48	2.62
CD	4.33	0.74	13.27	1.41	5.12	9.04

2. Effect of different levels of Manganese chelate 12% on yield of okra

Data presented in table 2 showed that effect of different levels of Manganese chelate1 2% on yield parameters of okra.

Regarding yield parameters, maximum length of fruits (10.0 cm), breadth of fruits (3.0 cm), average weight of fruit (22.5 g) was recorded in treatment consisted of Manganese chelate 12% of 1.0 g per litre. The highest yield per plot (16.2kg) and yield per hectare (120.3q/ha) was recorded in treatment consisted of Manganese chelate 12% of 1 of 1.0 g per litre in okra while lowest yield per hectare (115.5 q./ha) was recorded in control.

Table 2: Effect of different levels of Manganese chelate 12% on yield of okra

Treatments	Length of fruit (cm)	Breadth of fruit (cm)	Av. wt. of fruit (g)	Yield (Kg/plot)	Yield (qlha)
T1 Manganese Chelate 12% @ 0.5 g/l	9.4	3.0	20.5	15.8	117.7
T2 Manganese Chelate 12% @ 1.0 g/l	10.0	3.0	22.5	16.2	120.3
T3 Manganese Chelate 12% @ 1.5 g/l	9.8	2.9	22.2	16.0	119.5
T4 Control	9.0	2.9	19.7	15.1	115.5
SE	0.35	0.10	1.19	0.16	3.15
CD	1.21	0.36	4.10	0.55	10.89

3. Effect of different levels of Manganese chelate 12% on major pest and disease incidence of okra

Treatments	Powdery mildew PDI %
T1 Manganese Chelate 12% @ 0.5 g/l	3.60 (10.91)
T2 Manganese Chelate 12% @ 1.0 g/l	2.13 (8.38)
T3 Manganese Chelate 12% @ 1.5 g/l	3.09 (10.12)
T4 Control	3.62 (10.92)
SE ±	1.67
CD@5%	5.21
CV%	12.28

Data presented in table 3 showed that effect of different levels of Manganese chelate 12% on major disease and pest incidence of okra. The minimum powdery mildew in okra was observed in treatment with Manganese chelate 12% of 1.5 g per litre.

TRIAL REPORT 5

Name of the Organisation	: Acharya N.G. Ranga Agricultural University
Name of the Implementing Centre	: Regional Agricultural Research Station, A.N.G.R. Agricultural University, Lam, Guntur, AP
Name of the Sponsor	: M/s Aries Agrovet Industries
Product	: EDTA chelated Calcium
Crop and Season	: Chilly, Kharif 2003-2004
Design	: RBD
Replications	: Four
Spacing	: 56 X 30 cm
Treatments	: Six

Treatment Details

Sr. No	Particulars of the treatment
1	Recommended doses of NPK
2	Ca EDTA 0.5%
3	Ca EDTA 1.0 %
4	$CaNO_3$ 0.5%
5	$CaNO_3$ 1.0%
6	Control – Water Spray

Plot size	: 12.0 sq. m
Number of Sprays	: Three

1st Spray 50 days after transplanting
2nd Spray 75 days after transplanting
3rd Spray 100 days after transplanting

Results

Data on growth, yield characters and yield were recorded and are presented in Table 1.

Table 1 Effect of Ca-EDTA on plant growth, yield characters and yield in chilli

Treatments	Plant growth parameters			Fruit		No. of fruits / plant	Fruit weight / plant (g)	No. of seeds/ fruit	Dry fruit yield (q/ha)
	Height (cm)	Spread (cm)	No. of primary branches	Length (cm)	Girth (cm)				
T1	107.8	119.1	4.1	6.1	3.2	143.3	107.5	71.3	49.49
T2	112.5	117.3	4.0	6.2	3.3	159.8	119.8	70.2	49.29
T3	117.8	1220.3	4.3	6.6	3.3	169.8	127.4	69.8	56.97
T4	112.1	116.2	4.0	7.5	3.6	152.9	114.6	69.8	51.25
T5	106.0	112.4	3.6	6.9	3.3	162.8	122.1	70.0	48.56
T6	106.4	115.1	4.0	6.3	3.3	148.5	111.4	71.2	51.50
CD (P=0.05)	NS	NS	0.243	0.558	NS	9.453	7.086	NS	4.852

Effect of Ca-EDTA on growth characters

i) Plant height: The treatmental effects were not significant in respect of plant height. However, the plants sprayed with Ca EDTA 1% recorded the maximum plant height (117.8cm)

ii) Plant spread: The treatmental differences were not significant on plant spread. However, the plants sprayed with Ca EDTA 1% recorded the maximum plant spread (120.3cm)

iii) Number of primary branches: Significant differences were observed among the treatments. The plants sprayed with Ca EDTA 1% recorded the maximum number of primary branches (4.3)

Effect of chelacal on yield attributing characters and yield.

i) Fruit length: Among the treatments, the plants sprayed with CaN03 0.5% recorded the maximum fruit length (7.5cm) and significantly superior to all the other treatments.

ii) Fruit girth: The differences between the treatments were not significant in respect of fruit girth. However, the plants sprayed with $CaNO_3$ 0.5% recorded the maximum fruit girth (3.6cm)

iii) Number of fruits per plant: Highly significant differences were observed among the treatments with regard to number of fruits per plant. The plants sprayed with Ca EDTA 1.0% recorded the highest number of fruits per plant (169.8) and found significantly superior to all the other treatments evaluated.

iv) Fruit weight per plant: Highly significant differences were observed among the treatments with regard to the mean fruit weight per plant. The plants sprayed with Ca EDTA 1% recorded the highest mean fruit weight per plant (127.4g) and found significantly superior to all the other treatments but on par with CaNO3 1% (122.1g).

v) Number of seeds per fruit: Significant difference did not exist among the treatments in respect of number of seeds per fruit. However, the plants applied with recommended dose of fertilizers recorded the highest number of seeds per fruit (71.3)

vi) Dry chilli yield: The plants sprayed with Ca EDTA 1% recorded the highest dry chilli yield (56.97q/ha) and found significantly superior to all the other test treatments.

Conclusion:

Among the different treatments evaluated on Chilli, the plants sprayed thrice with Ca-EDTA 1% starting from 50th day after transplanting at 25 days interval recorded the highest dry chilli yield (56.97 q/ha) and found significantly superior to all the treatments tried. The number of fruits per plant and the mean fruit weight per plant were found to be the main contributory factors for the increased yield in the plants sprayed with Ca EDTA 1%.

TRIAL REPORT 6

Name of the Organisation	: Dr. Panjabrao Deshmukh Krishi, Vidyapeeth, Akola
Name of the Implementing Centre	: Agriculture Research Station, Washim, Dr.PDKV, Akola
Name of Agency	: Agro Input Manufacturers Association
Name of the product	: Calcium Chelate (6%).
Season	: Summer 2018
Design	: F.R.B.D
Replication	: Three
Plot Size	: Gross: 4.0 x 4.5 m^2 : Net 3.8 X 3.6 m^2
Spacing	: 4.0 x 3.0 m
Variety	: Bhagwa

Treatments : A. Calcium Level
: 1. Calcium Chelate at 0.5 gm per liter.
  2. Calcium Chelate at 1.0 gm per liter.
  3. Calcium Chelate at 1.5 gm per liter.
  4. Control (Water spr:ay).

: B. Application Stages
  1. Peak vegetative stage.
  2. 50% flowering stage.
  3. Fruit development stage.

Observation

Table 1: Growth and yield of pomegranate as influenced by different treatments.

Treatment	Plant height (cm)	Number of branches/plant	Calcium content in plant (%)	Fruit yield /plant (kg)
A. Calcium Level				
Calcium chelate at 0.5 gm per litre	176.3	12.7	0.79	25.8
Calcium chelate at 1.0 gm per litre	180.6	13.0	0.82	26.2
Calcium chelate at 1.5 gm per litre	182.2	13.6	0.83	26.9
Control (Water Spray)	171.1	12.4	0.69	23.5
S.E m±	3.85	0.20	0.03	1.1
CD at 5%	11.55	0.62	0.88	NS
B. Application Time				
Peak vegetative stage.	180.5	12.5	0.82	25.0
50% flowering stage.	177.3	12.9	0.83	25.4
Fruit development stage.	174.8	13.4	0.77	26.4
S.E m±	2.78	0.14	0.02	0.6
CD at 5%	8.34	0.44	NS	NS
C. Interaction				
CD at 5%	NS	NS	NS	NS

Results:

A. Growth parameters

1. Height of plant

Data on height of plant of pomegranate were influenced due to treatments of chelated ca and methods of application as presented in Table 1. The different concentration of chelated ca recorded increase in plant height over control. However, application of 1.5 g/lit chelated calcium (9%) was recorded maximum plant height and was at par with lower doses of calciu m (9%) i.e. 1.0g/lit and 0.5 g/lit, however, minimum increase in plant height was recorded in chelated calcium @ 0.5 g/lit. The data presented on time of application of chelated ca, the maximum increase in plant height was recorded at fruit development stage followed by at 50% flowering. However, minimum increase in plant height was recorded at vegetative stage.

2. Number of branches per plant

Data on number of branches of pomegranate were significantly influenced by different concentration of chelated ca and methods of application. The different concentration of chelated ca differed with respect to increase in number of branches of pomegranate. The treatment i.e. application of chelated calcium @ 1.5 g/lit at fruit development recorded maximum number of branches over lower dose of chelated calcium (9%). The data presented on method of application of chelated ca, the maximum increment in number of branches was recorded at vegetative stage over 50% flowering and fruit development stage.

3. Calcium content in plant (%)

Data on per cent calcium content in plant of pomegranate at harvest were significantly influenced by different concentration of chelated ca, however, did not influenced significantly with respect to methods of application. The different concentration of chelated ca differed with respect to per cent increase in calcium content in plant. The treatment i.e. application of chelated calcium @ 1.5 g/lit recorded higher calcium content in plant over lower dose of chelated calcium (9%). The data presented on method of application of chelated Ca, the maximum increase in per cent calcium content was recorded at fruit development stage over vegetative and at 50% flowering stage.

B. Yield

1. Fruit yield/plant (kg)

Foliar Ca fertilization had no significant effects on fruit yield per tree. The different concentration of Ca did not differed significantly with respect to increase in fruit yield of pomegranate. The treatment i.e. application of chelated calcium 1.5 g/lit recorded maximum fruit yield over lower dose of chelated calcium (9%). However, in case of method of application of chelated Ca (9%) the maximum increase in fruit yield of pomegranate was recorded at fruit development stage over foliar and at 50% flowering stage.

TRIAL REPORT 7

Name of the Organisation	: Vasantrao Naik Marathwada Krishi Vidyapeeth Parbhani
Department/location	: Sr. Research Officer, Horticulture Research Scheme (Pomology)
Name of the firm and address	: Agro Inputs Manufacturers Association (India) Chhatrapati Shivaji Market yard, Shop No.6, Ist Floor, Madhyavarti Bhavan, Gultekdi, Pune 411037.
Name of the product tested	: Calcium chelate (9%)
Year of conduct	: *Ambia bahar* 2015-16

MATERIAL AND METHODS

A product testing experiment for evaluation of calcium chelate (9%) on agronomic performance and yield in pomegranate var. Bhagwa was undertaken at Department of Horticulture, College of Agriculture, VNMKV, Parbhani during ambia bahar on 2.5 year old pomegranate to tree.

The experiment comprised of first factor A where four levels of calcium chelate (9%) *viz*. 0.5 gm per litre, 1.0 gm per litre,1.5 gm per litre and control (no application of calcium chelate 9%) combined with second factor B i.e method of application at three levels viz., first pray at peak vegetative growth, second spray at 50% flowering stage and third spray at fruit evelopment stage were tried involving twelve treatment combinations. The experiment was aid out in factorial randomized block design with three replication.Three plants were selected r each treatment under each replication for recoding vegetative growth parameter viz., plant fight and number of branches, flowering parameters viz., number of flower, number of male and rfect flower, while yield parameters viz., fruit drop and fruit set %, number of fruits at harvest, eld kg/plant and yield t/ha was recorded. All recommended agronomic practices were followed.

STATISTICAL ANALYSIS

The data recorded on above obervation was subjected for statistical analysis as suggested by Panse and Sukatme (1967).

RESULTS

A) Effect of different levels of calcium chelate 9% on growth, flowefing and yield of pomegranate.

It is evident from the table that different levels of calcium chelate 9% showed significant difference on vegetative growth, flowering and yield parameters of pomegranate.

Regarding vegetative parameters, significantly more plant height (5.1 cm) was recorded in treatment applied with calcium chelate 9% of l gm per litre while more number of branches per plant (1.16) was recorded in treatment of calcium chelate 9% of 1.5gm per litre.

In respect of flowering parameters, significantly more number of flowers per plant (110.5), nurnber of perfect flowers per plant (65.58), number of male flowers per plant (28.75) and sex ratio per plant (32.16) was recorded in treatment consisted of calcium chelate 9% of 1.5gm per litre.

Regarding yield parameters, significantly less fruit drop per plant (1.75 %) and more number of fruits at fruit set per plant (41.83), number of fruits at harvest per plant (40.16), yield per plant (11.90 kg) and yield per hectare (6.59 t) was recorded in treatment consisted of calcium chelate 9% of 1.5gm per litre.

B) Method application of calcium chelate 9% on growth, flowering and yield of pomegranate.

It is evident from the table that different methods of application of calcium chelate 9% showed significant difference on vegetative growth, flowering and yield parameters of pomegranaters except number of branches per plant, number of flowers per plant and number of male flowers per plant.

Regarding vegetative parameters, significantly more plant height (5.12 cm) was recorded in treatment applied with calcium chelate 9% at fruit development stage.

In respect of flowering parameters, significantly more number of perfect flowers per plant(68.25) and sex ratio per plant (34.34) was recorded in treatment consisted of calcium chelate 9% at fruit development stage.

Regarding yield parameters, significantly less fruit drop per plant (1.81 %) and more number of fruits at fruit set per plant (43.12), number of fruits at harvest per plant (42.12), yield per plant (9.15 kg) and yield per hectare (6.01 t) was recorded in treatment consisted of calcium chelate 9% at fruit development stage.

C) Interaction effect of different levels and methods application of calcium chelate 9% on growth, flowering and yield of pomegranate

It is evident from the table that different levels and methods of application of calcium chelate 9% showed significant difference on vegetative growth, flowering and yield parameters of pomegranate except number of branches per plant, and number of male flowers per plant.

Regarding vegetative parameters, significantly more plant height per plant (6.75 cm) was recorded in treatment combination consisting of calcium chelate 9% of 1.5gm per litre at fruit development stage.

In respect of flowering parameters, significantly more number of flower per plant (135.87), number of perfect flowers per plant (81.75) and sex ratio per plant (40.75.34) was recorded in treatment combination consisting of calcium chelate 9% of 0.5gm per litre at fruit development stage.

Regarding yield parameters, significantly less fruit drop per plant (1.00 %) was recorded in treatment combination consisting of calcium chelate 9% of 1.5gm per litre at 50% flowering stage and more number of fruits at fruit set per plant (53.75), number of fruits at harvest per plant (48.50), yield per. plant (12.05 kg) and yield per hectare (6.9 t) was recorded in treatment combination consisting of calcium chelate 9% of 1.5 gm per liter at fruit development stage.

Table Field evaluation of efficacy of cacium chelate (9%) on growth flowering and yield of pomegranate var. Bhagwa.

Treatment details	Growth Parameter		Flowering Parameter				Yield Parameter				
	Plant height	No.of branches/ plant	No.of flowers per plant	No.of perfect flowers per plant	No.of male flowers per plant	Sex ratio	No.of fruits at fruit set per plant	No.of fruits drop per plant	No.of fruits at harvest per plant	Yield kg / plant	Yield t/ha
Factor A means											
Ti	4.66	0.91	128.87	62.08	27.42	30.16	37.00	2.83	34.33	8.03	4.99
T2	5.16	0.75	104.75	62.75	26.16	30.12	39.75	2.12	37.50	9.10	5.53
T3	4.75	1.16	110.50	65.58	28.75	32.16	41.83	1.75	40.16	11.90	6.59
T4	2.75	1.00	87.50	40.00	21.00	21.00	23.50	1.50	24.00	7.87	4.86
SE +	0.56	0.22	4.90	1.74	2.49	0.78	1.72	0.16	1.80	0.12	0.15
CD at 5%	1.68	0.66	14.72	5.24	7.47	2.36	5.16	0.49	5.41	0.37	0.49
Factor B means											
B1	4.18	0.81	109.37	59.55	25.06	28.81	37.18	2.00	35.81	9.05	5.48
B2	3.68	1.00	102.62	45.00	26.56	21.93	26.25	2.34	24.06	8.91	5.30
B3	5..12	1.06	111.72	68.25	25.87	34.34	43.12	1.81	42.12	9.28	6.01
SE +	0.48	0.19	4.24	1.51	2.15	0.68	1.49	0.14	1.56	0.10	0.15
CD at 5%	1.45	NS	NS	4.54	NS	2.04	4.47	0.43	4.70	0.32	0.48
A X B means											
T1B1	4.00	0.50	125.75	55.00	27.00	26.25	37.00	2.50	34.50	8.00	5.11
T1B2	4.00	0.75	125.00	49.50	28.25	23.50	28.50	3.75	26.35	7.95	4.85
T1B3	6.00	1.50	135.87	81.75	27.00	40.75	45.50	2.25	43.75	8.35	5.24
T2B1	6.25	0.75	113.50	69.00	27.50	30.87	41.50	1.75	39.50	8.45	5.60
T2B2	4.25	1.00	93.00	41.50	25.75	20.75	24.00	3.12	26.75	8.22	5.40
T2B3	5.00	0.50	107.75	77.75	25.25	38.75	49.75	1.50	52.25	9.25	5.80
T3B1	3.75	1.00	110.75	74.25	24.75	37.12	46.75	2.25	45.25	11.95	6.70
T3B2	3.75	1.25	105.00	49.00	31.25	22.50	29.00	1.00	26.75	11.65	6.10
T3B3	6.75	1.25	115.75	73.50	30.25	36.87	53.75	2.00	48.50	12.05	6.90
T4B1	2.75	1.00	87.50	40.00	21.00	21.00	23.50	1.50	24.00	7.87	4.85
2T4B2	2.75	1.00	87.50	40.00	21.00	21.00	23.50	1.50	24.00	7.87	4.85
T4B3	2.75	1.00	87.50	40.00	21.00	21.00	23.50	1.50	24.00	7.87	4.85
SE +	0.97	0.38	8.49	3.03	4.31	1.36	2.98	0.28	3.12	0.21	0.28
CD at 5%	2.92	NS	25.481	9.09	NS	4.09	8.94	0.86	9.39	0.65	0.85

TRIAL REPORT 8

Name of the Organisation	: Mahatma Phule Krishi Vidyapeeth, Rahuri
Name of the Implementing Centre	: AICRP on Arid Zone Fruits, Department of Horticulture, Mahatma Phule Krishi Vidyapeeth, Rahuri-413272, Dist Ahmednagar (M.S.)
Name of the Company	: Agro Manufacturers Association (India)
Product Name	: Calcium Chelate (9%)
Crop	: Pomegranate
Variety	: Phule Bhagwa Super
Season and year	: 2015 (*Ambia bahar*)
Spacing	: 4.5 x 3.0 m
Plant unit	: 5
Design	: Randomized Block Design (RBD)
Number of treatments	: Four (04)
Replications	: Five (05)
Start of *bahar*	: 15-02- 2015
Date of harvesting	: 20-08-2015 to 30-09-2015
Application method	: Foliar spray

Table 1: Treatment details

SN	Treatments	Treatment Details
1	T_1	RDF + Calcium Chelate @ 0.5 g per liter
2	T_2	RDF + Calcium Chelate @ 1.0 g per liter
3	T_3	RDF + Calcium Chelate @ 1.5 g per liter
4	T_4	RDF + No spray (Control)

Spraying Schedule: Date of Spraying & Product

SN	Spray	Stage of spray	Date of spray
1	I[st] Spray	During peak vegetative growth	23-02-2015
2	IInd Spray	50 % flowering stage	13-03-2015
3	IIIrd Spray	Fruit development stage	16-05-2015

Observations recorded

A) Plant characters

    1 Plant height(m)
    2 No. of branches
    3 No. of flowers

B) Fruit Characters

    1 Av. weight of fruit (g)
    2 Fruit droP (%)
    3 Fruit cracking (%)
    4 Calcium (%)
    5 Total soluble solids ('B)
    6 Acidity (%)

C) Yield Characters

    1 Average number of fruits/Plant
    2 Yield (kg/Plant)
    3 Yield (t/ha)

D) Methods used for biochemical analysis

Micronutrients	Method used for Analysis
Calcium	Atomic Adsorption Spectrophotometer (A.O. A'O 1990)

Results

The trial was conducted in the 2015 *Ambe bahar* as per protocol supplied by the company. The periodical observations were taken according to the growth of the plant and details are given in the respective tables'

A) Growth Parameters

The growth of orchard was satisfactory throughout the season. Though the results were non-significant the maximum plant height was noticed in treatment T-3 (RDF + Calcium Chelate @l.5 g per litre)i.e 1.74 m followed byT-2(RDF+CalciumChelate@ 1.0g per litre) 1 .70 m. The maximum nurnber of branches were found in treatment T-4 i.e. 4.60 though the results are non-significant. Significantly maximum number of female flowers were observed in treatment T-2 (90.60 flowers / plant) followed by treatment T-3 (85.00 flowers / plant).While the % fruit setting was maximum in T-2 (93.41 %) followed by treatment T-1 (RDF + Calcium Chelate@ 0.5 g per liter).

B) Yield Parameters

Among the treatments the maximum results were significant except the plant height and no. of branches per plant. The highest number of fruits per plant were registered by treatment T-2 i.e.(82 fruits/plant) followed by treatment T-3 (79.40 fruits /plant), while it was recorded lowest by treatment T-4 i.e. 67.40 fruits/plant (RDF + No spray) i.e. control. The average weight of fruit was significantly noticed maximum by treatmentT-2(302.00 g) followed by treatment T- 3 (286.40 g). The highest fruit yield was recorded by the treatment T-2 (22.65 kg/plant and 16.76 t/ha) followed by treatment T-3 i.e. (21.47 kg/plant and 15.88 t/ha). The scorching effect on leaves and fruits was observed in the treatment T-3.

C) Quality parameters

The maximum total soluble solids were recorded by the treatment T-3 (13.96%) followed by treatment T-2 (13.80 %) while lower acidity was noticed in treatment T-3 (0.32%).

D) Biochemical parameters

No significant differences were observed within the treatment in respect of Calcium content, but treatment T-3 recorded maximum (9.7 mg/100g) followed by treatment T-1 and T-2 i.e. (9.5 mg/100g).

The negligible fruit drop in percentage was observed in all the treatments. Treatment T-3 recorded 0.64 percent fruit drop which was minimum amongst all the treatments. No fruit cracking was observed in any of the treatments.

Table 1: Fruit, plant and yield characters of pomegranate

Sr. No		1	2	3	4		
Treatments		T1	T2	T3	T4-control	S.E±	CD at 5%
Plant Height (m)		1.69	1.7	1.74	1.69	0.03	NS
No. of branches		4	4.4	4.4	4.6	0.43	NS
No.of flowers / plant	Male	101.2	127.4	136.2	95.8	9.28	27.95
	Female	79.4	90.6	85	73.4	5.01	15.08
Av. Weight of fruit	(g)	264	302	286.4	262.4	12.08	36.4
No.of fruits / plant		74.6	82	79.4	67.4	4.33	13.03
% fruit set		93	93.41	90.5	91.82		
Yield/plant	(kg)	20.8	22.69	21.47	20.41	0.56	1.68
Yield	(t/ha)	15.3	16.76	15.88	15.1	0.41	1.24
TSS	(%)	13.72	13.8	13.96	13.69	0.29	NS
Acidity	(%)	0.33	0.33	0.32	0.33	0.01	NS
Calcium	(mg/100g)	9.5	9.5	9.7	9.4		
Fruit drop		1.74	0.64	0.88	1.8	0.38	1.13
Fruit cracking		0	0	0	0		
Scorching on leaf and fruits				observed			

Conclusion:

Application of GRDF (625: 250 :250 g/plant+30k g FYM) along with 3 sprays of Calcium Cheated 9 % @ 1.0 g/liter (treatment T -2) recorded maximum % fruit setting (93.41%), average weight of fruit (302.00 g) and yield (22.65 kg/plant) and (16.76) t/ha followed by treatment T -3.

TRIAL REPORT 9

Name of Organisation	: Vasantrao Naik Marathwada Krishi Vidyapeeth, Parbhani
Location of Testing	: Horticulture Research Scheme (Vegetable) Sub Campus, VNMKV, Parbhani-43t 402
Product detail	: Copper Chelate (12%)
	Copper Chelate 12% @0.5g/L
	Copper Chelate 12% @1.0g/L
	Copper Chelate 12% @1.5g/L
Name of Firm	: Agro Inputs Manufacturers Association (India) Chhatrapati Shivaji Market Yard, Shop No.6,1 St Floor, Madhyavarti Bhavan, Gultekdi, Pune 411 037.
Test Season	: Kharif 2015-2016

Material and Methods:

A product testing experiment for evaluation of Copper Chelate (12%) on agronomic performance and yield in onion var. Parbhani Bhendi (PBN-OK-1) was undertaken at Horticulture Research Scheme (Vegetable), Sub Campus, VNMKV, Parbhan, VNMKV, during *kharif* 2015-2016. The details of material used and methodology adopted is given as below.

a) Variety	: Onion- Phule Samarth
b) No. of Treatments	: Four
c) Treatment Detail:	
	T1- Copper Chelate 12% @0.5g/L
	T2- Copper Chelate 12% @1.0g/L
	T3- Copper Chelate 12% @1.5g/1_,
	T4- Control
d) No. of Replications	: Five
e) Experimental Design	: Randomized Block Design
f) Plot Size	: 3.1 X 1.5 m
g) Spacing	: 30 X 15 cm
h) Date of Seed sowing	: 27.07.2015
i) Date of transplanting	: 24.8.2015

i) Soil Type : Medium black cotton soil
j) Fertilizer dose : 120:60:60

Application Method:

First Spray:-21-30 Days after sowing/transplanting (peak growth)
Second spray: - Tillering stage
Third Spray: - Bulb development stage

Result:

1. Effect of different levels of Copper chelate 12% on growth of onion

It is evident from the table 1 that different levels of Copper chelate 12% showed difference on vegetative growth parameters of onion.

Regarding vegetative parameters, maximum plant height (47.8 cm) was recorded in treatment applied with Copper chelate 12% of 1.5g per litre while more number of leaves per plant 10.6) was recorded in treatment of Copper chelate 12% of 1.0g per litre.

Table 1: Effect of different levels of Copper chelate 12% on growth and yield of onion

Treatments	Plant height (cm)	No.of leaves	Length of bulb (cm)	Diameter of bulb (cm)	Avg. wt. of bulb (g)	Yield/ plot (kg)	Yield (q/ha)
T1- Copper chelate 12% @ 0.5 g/l	45.8	9.8	7.76	21.27	140.5	10.7	270.4
T2- Copper chelate 12% @ 1.0 g/l	47.2	10.6	9.32	24.2	151.9	12.7	302.1
T3- Copper chelate 12% @ 1.5 g/l	47.8	10.5	8.51	23.3	149.3	11.2	288.9
T4- Control	44.2	9.2	7.16	20.82	136.6	9.5	248.7
SE	1.40	0.49	0.36	0.19	1.83	0.57	19.84
CD	N/A	N/A	N/A	N/A	6.45	2.01	70.00

Regarding yield parameters, maximum length of bulb 9.32 cm), diameter of bulb (24.2 cm), average weight of fruit (151.9 g) was recorded in treatment consisted of Copper chelate 12% of 1.0g per litre. The highest yield per plot (12.7 kg) and yield per hectare (302.1 q/ha) was recorded in treatment consisted of Copper chelate 12% of 1.0 g per litre in onion while lowest yield per hectare (248.7 q/ha) was recorded in control

2. Effect of different levels of Copper chelate 12% on major pest and disease incidence of onion

Treatments	Purple Blotch PDI %
T1- Copper chelate 12% @ 0.5 g/l	4.34 (11.97)
T2- Copper chelate 12% @ 1.0 g/l	2.97 (9.87)
T3- Copper chelate 12% @ 1.5 g/l	2.34 (8.75)
T4- Control	6.07 (14.16)
SE±	1.05
CD @5%	3.09
CV%	12.86

Data presented in table 2 showed that effect of different levels of Copper chelate 12% on major disease and pest incidence of onion. The minimum purple blotch in onion was observed in treatment with Copper chelate 12% of 1.5 g per litre. Regarding vegetative parameters, maximum plant height (47.8 cm) was recorded in treatment applied with Copper chelate 12% of 1.5g per litre while more number of leaves per plant 10.6) was recorded in treatment of Copper chelate 12% of 1.0g per litre.

TRIAL REPORT 10

Name of Organisation	: Dr. Panjabrao Deshmukh Krishi Vidyapeeth, Akola
Department/ Section	: Chilli & Vegetable Research Unit, Dr. PDKV, Akola
Name of Product	: Copper chelate (12%)
Name of Firm	: Agro Inputs Manufacturers Association

Technical Program

a) Experimental Details

  i) Progressive year : 2017-18
  ii) Design : FRBD
  iii) Replications : Five
  iv) Plot size : Gross: 3.2 m X 1.2 m.
        Net: 3.00mX1.0m.
  v) Spacing : 10 cm X 10 cm
  viii) Variety : All Green
  viii) Treatrnents : Factor A   Factor B
        D_1- 21st February C_1- 0.5 g/lit.
        D_2- 2nd March C_2- 1 g/lit.
        D_2- 12th March C_3 – 1.5 g/lit.
                  C_4 – Water Spray

Treatment Details

D_1T_1	: Copper chelate (12%) @ 0.5 g/lit as at 2l-30 DAT
D_1T_2	: Copper chelate (12%) @ 1.0 g/lit as at 35-40 DAT
D_1T_3	: Copper chelate (12%) @ 1.5 g/lit as at 45-50 DAT
D_1T_4	: Water Spray
D_2T_1	: Copper chelate (12%) @ 0.5 g/lit as at 2l-30 DAT
D_2T_2	: Copper chelate (12%) @ 1.0 g/lit as at 35-40 DAT
D_2T_3	: Copper chelate (12%) @ 1.5 g/lit as at 45-50 DAT

D_2T_4 : Water Spray
D_3T_1 : Copper chelate (12%) @ 0.5 g/lit as at 21-30 DAT
D_3T_2 : Copper chelate (12%) @ 1.0 g/lit as at 35-40 DAT
D_3T_3 : Copper chelate (12%) @ 1.5 g/lit as at 45-50 DAT
D_3T_4 : Water Spray

Initial status of experimental soil

pH		7.64
EC (dS/m)	:	0.15
Organic carbon (gm kg-1)	:	4.21
Available N (kg ha-1)	:	220
Available P (kg ha-1)	:	8.75
Available K (kg ha-1)	:	318
Available S (mg kg-1)	:	9.48

Observations to be recorded

Sr. 'No.	Characters	Observations
1.	Growth attributes	_Plant height (cm)
		No. of leaves bulb^{-1}
		size of bulb-polar (cm)
		size of bulb- equatorial (cm)
		Neck thickness (cm)
2	Yield and yield attributes	Yield plot"' (kg)
		Yield ha^{-1} (q.)
3	Observations related to quality parameter	TSS 'Bx

Material and Methods:

The experiment was conducted at research field of CVRU., Dr. PDKV., Akola during rabi season 2017-18. The trial was conducted in FRBD with 2 factors i.e. Dates of application of 12 % copper chelate with three levels i.e. 21st February (D_1), 2nd March (D2) and 12th March (D3) as sub factors. Whereas, concentration of copper chelate (12%) as a main factor with 4 levels i.e. C1-0.5 gait, C2-1 g/lit, C3-1.5 eft C4-Water spray.

The application was made as per the treatments on every plot as per the plan of layout. Different observations were undertaken as per the protocol of company and data was recorded accordingly. The statistical analysis was undertaken with the help of methods and experimental design suggested by Panse and Sukhatme (1967).

Results:

From the various data presented in following tables on vegetative and qualitative characters of onion i.e. Plant height (cm), No. of leaves per bulb, size of bulb-polar (cm), size of bulb- equatorial (cm), Neck thickness (cm), and yield q/ha are described with appropriate headings and sub headings.

Plant Height (cm):

Effect of concentrations (C)

It is evident from the data presented in Table 1 that, significantly the maximum (75.35 cm) plant height of onion was observed due to an applications of 1.0 g per liter copper chelate. However, it was minimum (63.50cm) with control treatment i.e. water spray.

Effect of spraying days interval (D)

Significantly the maximum (80.54 cm) plant height of onion plant was produced when the copper chelate was applied at 30 DAS as compared to rest of the levels of spraying days interval. Minimum plant height (57.17 cm) of onion was noticed with application of copper chelate at 30 DAS.

Furthermore, the interaction effect of different concentrations of copper chelate and spraying was found statistically non-significant.

Leaves per plant: -

Effect of concentration:

From the data presented in table 1, it is revealed that, significantly the maximum (14.04) leaves per plant to the onion plant were produced due to an application of copper chelate @ concentrations of 1.0 WM of water. The effects of concentrations 0.5 g/lit (C1) of water and 1.5 g/lit of water (C2) were found statically at par with each other. Whereas, significantly the minimum (10.47) leaves per plant to onion were obtained with control (water spray) level of the concentration (C4).

Effect of spraying days interval: -

Significantly the maximum (14.11) leaves per plant were produced to the onion plant getting copper chelate application at 30 DAS. However, it was significantly minimum (10.56) with later (60 DAS) spray of copper chelate in the present investigation.

The interaction effect between concentration and spraying days interval of copper chelate was found to be statistically non-significant.

Yield contributing characters:

Size of onion bulb polar and equatorial concentration: -

Data presented in table expressed that, significantly the maximum (4.16 cm and 5.13 cm) polar and equatorial diameter of onion bulbs were recorded with application of copper chelate @ 1.0 gait (C2) on onion plants. Whereas, these were recorded minimum (3.87 cm and 1.18 cm, respectively) with an application of water spray treatment.

Spraying days after transplanting:

As far as the size of onion bulbs produced due to an application of copper chelate at 30 DAS, significantly the maximum polar and equatorial diameter (3.97 and 4.90 cm, respectively) was obtained in this experiment. However, it was minimum (3.30 and 4.22 cm, respectively) with copper chelate application at 60 DAS.

The interaction effects between the concentration and spraying days interval does not show any significant differences as far as the size (polar and equatorial) of onion bulb are concerned.

Yield of onion bulbs (q/ha):

Levels of concentrations:

From the data presented in table 1, it is observed that, significantly the maximum (279.19 q/ha) onion bulb were produced due to an application of 1.0 g/lit of copper chelate. Whereas, significantly the minimum (269.89 q/ha) onion bulb were obtained without application of copper chelate (water spray). The growth regulator stimulated the plant growth and bulb development which has given more opportunity for carbohydrate assimilation needed for better yield. Similar results were also reported by Bose et al (2009) and M.A. Quadir et al (2014) in onion under Jabalpur (MP) and Lahore (Pakistan) conditions.

Levels of spraying days after transplanting:

There was significant differences observed due to various levels of spray of copper chelate after 30, 45 and 60 DAS. Significantly the maximum (280.45 q/ha) onion bulbs were produced due to first level (30 DAS) of copper chelate application. However, it was observed to be minimum (267.76 q/ha) with 60 DAS level of application of spraying days after transplanting of onion seedlings.

Interaction effect:

It is evident from the data presented in table 1 that, the interaction between different concentration of copper chelate and its spraying interval after transplanting was found statistically significant. The maximum (295.92 q/ha) onion bulbs was produced due to an application of copper chelate @ 1.0 gat at 30 DAT (C2D1). The treatment combinations other than C4D3 were found statistically at par with each other. However, significantly minimum (254.89 q/ha) of onion bulb yield was produced due to application of water spray at 60 DAT in the present investigation (Table 2). The weight of seed individually as well as test weight was increased in the copper chelate (12%) and other micro nutrients applied plots, than that of control treatment plot, this might be due to the fact that, micronutrient enhances the photosynthesis rate due to which more food was formed and stored in bulbs. These results are in line with the findings of Muhammad A Quadir et al (2014) under Lahore (Pakistan) Conditions.

Table 1: Effect of different concentration of 12% copper Chelate on growth and yield in onion

Treatment	Plant height (cm)	No.of leaves plant^{-1}	Size of bulb polar(cm)	Size of bulb – equatorial(cm)	Neck thickness (cm)	TSS ^{0}Bx	Yield ha^{-1}(q.)
Concentration							
C_1- 0.5g/lit	65.86	11.55	3.39	4.33	1.21	11.55	274.63
C_2- 1.0g/lit	75.35	14.04	4.16	5.13	1.29	13.46	279.19
C_3-1.5g/lit	68.59	12.85	3.83	4.83	1.22	12.25	272.09
C_4- water spray	63.50	10.47	3.10	3.87	1.18	10.54	269.89
F test	Sig.	Sig.	Sig.	Sig.	NS	NS	Sig.
SE(m)+	1.59	0.52	0.10	0.10	0.014	0.52	2.61
CD at 5%	4.58	1.50	0.29	0.30	-	-	7.55
Spraying Days Interval after transplanting							

D_1 - 30 DAT	80.54	14.11	3.97	4.90	1.29	13.29	280.45
D_2 - 45 DAT	68.79	12.02	3.60	4.50	1.21	12.02	273.61
D_3 - 60 DAT	57.17	10.56	3.30	4.22	1.17	10.56	267.76
F test	Sig.	Sig.	Sig.	Sig.	NS	Ns	Sig.
SE(m)+	1.37	0.45	0.087	0.09	0.012	0.45	1.87
CD at 5%	3.97	1.30	0.25	0.26	-	-	5.43
Interaction (CXD)							
F test	NS	NS	NS	NS	NS	NS	Sig.
SE(m)+	2.75	0.90	0.17	0.018	0.024	0.89	3.75
CD at 5%	-	-	-	-	-	-	10.86

Table 2: Interaction effect between the levels of concentration and levels of spraying days interval after transplanting due to copper chelate (12%) on onion

Treatment	D_1	D_2	D_3	Mean
C_1- 0.5g/lit	272.92	276.92	274.63	274.63
C_2- 1.0g/lit	295.92	272.34	269.32	279.193
C_3-1.5g/lit	274.63	269.32	272.34	272.097
C_4- water spray	278.92	275.878	254.895	269.898
Mean	280.453	273.614	267.796	
F test				Sig.
SE(m)+				3.76
CD at 5%				10.86

Conclusions: -

From one season data collected in the present investigation following conclusions could be drawn.

1. Significant influence of copper chelate on plant height, leaves per plant and bulb yield in onion were observed due to application of 1.0 Oil concentration of spra as compared to rest of the levels of concentrations
2. As far as the time of application of copper chelate on onion crop after transplanting are concerned, significantly the superior plant height, leaves

per plant and yield of onion bulbs along with its size were reported at 30 DAT as against 45 DAT and 60 DAT.
3. Interaction effect between concentrations and time of application of copper chelate (C X D) showed statistically significant differences only with yield of onion bulbs. However, rest of vegetative and qualitative growth observations doesn't showed significant differences in the present investigation.

Such, above results/conclusion is based on observations of only one season, further study would be more useful to draw concrete outcome.

TRIAL REPORT 11

Name of the Organisation	: Dr. Panjabrao Deshmukh Krishi, Vidyapeeth, Krishi Nagar, Akola
Name of the Implementing Centre	: Agriculture Research Station, Sonapur. Gadchiroli
Name of the product	: Magnesium Chelate (6%).
Name of Firm	: Agro Inputs Manufacturers Association
Specific area	: Foliar spray
Treatments (Foliar application)	: a. Magnesium Chelate at 0.5 gm per liter b. Magnesium Chelate at 1J-.) gm per liter c. Magnesium Chelate at 1.5 gm per liter d. Control
Replications	: 5 (five)
Methodology	
Experimental Design	: Randomized block design (RBD)
Variety	: Pusa Ruby
Plot size	: 5.4 m X 3.75 m.----- 20.25 sq.m
Spacing	: 75 x 60 cm
Treatment	: 0.5. 1.0 and 1.5 gm per liter
Number of Sprays	: 21-30 days after transplanting. 50% flowering stage, and fruit development stage
Observations recorded	: plant height, number of branches per plant number of flowers per cluster number of fruits per cluster number of fruits per plant. average fruit weight (g) fruit length (cm) fruit girth (cm) and yield per hectare in quintals.
Fertilizer application	: 75 to 100 : 60 : 50 kg NPK ha

Table 1: Growth, quality and yield parameter of tomato to different levels of Magnesium concentration, the data recorded at Agriculture Research Station, Sonapur, Gadchiroli during kharif season of 2018-19

Treatment	Plant height (cm)	No.of branches per plant	No.of flower per cluster	No.of fruits clusters	Average fruit weight (gm)	Fruit length (cm)	fruit girth (cm)	No.of fruits per plant	Fruit Yield plant'	Fruit yield (kg/plot)	Fruit yield (q/ha)
		2	3		5			8		10	11
1. Mg at 0.5 gm per liter	109.25	6.25	3.25	2.25	36.75	5.80	3.00	34.75	5.00	37.03	182.84
2. Mg at 1.0 gm per lite	129.38	7.75	3.75	3.25	40.50	7.13	3.38	41.25	5.88	41.13	203,09
3. Mg at 1.5 gm per lite	113.25	6.75	3.25	2.25	35.00	5.90	3.25	34.00	4.32	36.80	181.73
4. Control	107.50	4.50	2.50	1.75	32.75	5.55	3.00	28.00	4.10	36.17	178.52
F'test											Sig.
S.E. (m)		1.0									1.202
CD at 5%											3.743
CV											7.112

Results

Among the different levels of Mg (6%) chelate, 1.0 g per litre level showed significant increase in plant height (129.38 cm), number of branches (7.75), number of flower cluster (3.75), number of fruits cluster (3.25), number of fruits per plant (41.25), average fruit weight (40.50 gm), fruit length (7.13 cm), fruit girth (3.38) and yield per ha (203.09 q). Based on above result, it is recommended that 6% Mg concentration @1.0 g per litre of water should be applied to tomato for better growth and yield.

TRIAL REPORT 12

Name of the Organisation : Mahatma Phule Krishi Vidyapeeth, Rahuri
Name of the Implementing Centre : Tomato Improvement Scheme, Department of Horticulture, MPKV, Rahuri

Name of the product : Magnesium Chelate (6%).

Name of firm : Agro Inputs Manufacturers Association

Experimental details

Variety/hybrid : Phule Raja
Season and year : Rabi 2013-14
Design : Randomized Block Design (RBD)
Replications : Four
No. of treatments : Five
Plot size : 3.6 x 3.3 m^2
Spacing : 90 x 30 cm

Treatment details:

Treatments	Product	Quantity
T1	Magnesium Chelate (6%)	0.5g/ lit
T2	Magnesium Chelate (6%)	1.09g/ lit
T3	Magnesium Chelate (6%)	1.5g/ lit
T4	Control (water spray)	--

No. of sprays and time of application after transplanting

1" spray	2nd spray	3rd spray
21-30 days after transplanting	50% Flowering stage	Fruit development stage
19/01/2014	27/01/2014	2/02/2014

Methodology

Recommended dose of fertilizer as 300:150:150 NPK Kg/ha + FYM -20 t/ha was applied to all the treatments.

The products including Magnesium Chelate (6%) were sprayed after transplanting at specific intervals as given above.

Observations recorded   : Plant height (cm)
                             Number of branches/plant
                             Average fruit weight (g)
                             Number of fruits/plant
                             Fruit yield /plant (kg)
                             Fruit yield /ha (t/ha)

Table 1: Effect of Magnesium Chelate (6%) on growth and yield in tomato during Rabi 2013-14

Treatments	Height of plant (cm)	Number of branches/ plant	Average weight of fruit (g)	Average number of fruits/plant	Yield / plant (kg)	Yield/ ha (t)
Ti	136.66	6.12	85.68	31.45	1.82	61.24
T2	144.31	6.24	87.83	32.40	1.93	65.14
T3	127.44	5.88	83.88	27.65	1.78	59.96
T4	123.17	5.88	80.88	24.33	1.62	54.60
SE±	1.02	0.36	0.57	0.69	0.02	--
CD at 5 %	2.23	0.79	1.26	1.50	0.05	--

Results:

It was observed that the yield per ha. was differed significantly due to various spray treatments. The maximum yield per hectare was recorded in treatment T2 (65.14 tons/ha) and lowest yield was revealed in treatment T4 (54.60 tons/ha) i.e. control. The average number of fruits per plant was highest in the treatment T2 (32.40) followed by T1 (31.45). The average fruit weight was recorded highest in the treatment T1 (87.83 g). The average yield per plant was maximum in the treatment T2 (1.93kg). The plant height (144.31cm) and the number of branches per plant (6.24) were highest in the treatment T2.

The overall results indicated that, the treatment T2 is found superior for higher yield over control.

TRIAL REPORT 13

Name of Organisation	: Dr. Balasaheb Sawant Konkan Krishi Vidyaoeeth Dapoli 415712, .Dist. Ratnagiri (M.S.)
Department/Section	: Agricultural Research Station, Phondaghat, Tal: Kankavli. Dist: Sindhudurg (MS) — 416 601
Name of the research centre	Agricultural Research Station, Phondaghat, Tal: kankavali, Dist:Sindhudurga(MS) - 416601
Latitude	16^0-22'35"
Longitude	73^0-47'18'
Elevation	165.11m
Soil Type	Sandy Loam
Soil pH	5.6-6.5
Soil Texture	Clay Loam

Months — MONTHLY WEATHER DATA-2015-16 (RABI SEASON)

Parameter	Dec	Jan	Feb	Mar	Apr	Total	Range
No. of Rainy days	0	0	0	0	0	0	-
Total Rainfall (mm)	0	0	0	0	0	0	-
Avg. Max. Temp.(^{0}C)	NA	NA	NA	NA	NA	NA	-
Avg. Mini. Temp. (^{0}C)	NA	NA	NA	NA	NA	NA	-
Humidity %	NA	NA	NA	NA	NA	NA	-
*Solar Radiation(Cal/cm^2)	NA	NA	NA	NA	NA	NA	-
*Avg. Sunshine Hours	NA	NA	NA	NA	NA	NA	-
*Avg, Wind Velocity(Km/h)	NA	NA	NA	NA	NA	NA	-

*Per day based on average over each month. NA: Not Available

Week No.	Metrological Period	Rainfall (mm)	Rainy Days	Max. Temp ^0C	Min. Temp ^0C	RH (%)
49	03-12-15 to 09-12-15	0.0	0.0	NA	NA	NA
50	10-12-15 to 16-12-15	0.0	0.0	NA	NA	NA
51	17-12-15 to 23-12-15	0.0	0.0	NA	NA	NA
52	24-12-15 to 31-12-15	0.0	0.0	NA	NA	NA
01	01-01-16 to 07-01-16	0.0	0.0	NA	NA	NA
02	08-01-16 to 14-01-16	0.0	0.0	NA	NA	NA
03	15-01-16 to 21-01-16	0.0	0.0	NA	NA	NA
04	22-01-16 to 28-01-16	0.0	0.0	NA	NA	NA
05	29-01-16 to 04-02-16	0.0	0.0	NA	NA	NA
06	05-02-16 to 11-05-16	0.0	0.0	NA	NA	NA
07	12-02-16 to 18-02-16	0.0	0.0	NA	NA	NA
08	19-02-16 to 25-02-16	0.0	0.0	NA	NA	NA
09	26-02-16 to 04-03-16	0.0	0.0	NA	NA	NA
10	05-03-16 to 11-03-16	0.0	0.0	NA	NA	NA
11	12-03-16 to 18-03-16	0.0	0.0	NA	NA	NA
12	19-03-16 to 25-03-16	0.0	0.0	NA	NA	NA
13	26-03-16 to 01-04-16	0.0	0.0	NA	NA	NA
14	02-04-16 to 08-04-16	0.0	0.0	NA	NA	NA
15	09-04-16 to 15-04-16	0.0	0.0	NA	NA	NA
16	16-04-16 to 22-04-16	0.0	0.0	NA	NA	NA
17	23-04-16 to 29-04-16	0.0	0.0	NA	NA	NA
18	30-04-16 to 06-05-16	0.0	0.0	NA	NA	NA

Sponsor Organization : Agro Inputs Manufacturers Association (India) Chhatrapati Shivaji Market Yard, Shop No.6,1 St Floor, Madhyavarti Bhavan, Gultekdi, Pune 411 037.

General information

a) Location : Agricultural Research Station, Phondaghat, Tal: Kankavli. Dist: Sindhudurg (MS) 416 601

b) Crop and Variety : Crop: Tomato, Variety: Arka Alok

c) Design/Layout : Randomized Block Design

d) Replications : Five

e) Treatments	: T1 – Control
	T2- Spraying of Magnesium chelate powder 6% @ 0.5 gm / liter of water
	T3- Spraying of Magnesium chelate powder 6% @ 1.0 gm / liter of water
	T4 - Spraying of Magnesium chelate powder 6% @ 1.5 gm / liter of water
	T5 - Spraying of Magnesium chelate powder 6% @ 2.0 gm / liter of water
f) Application method	: 1st spray - 21 – 30 days after sowing / t transplanting (at peak vegetative growth)
	2nd spray – At 50% flowering period
	3rd spray – At fruit development stage
g) Plot Size	: Gross:5.40mX4.80 m
	Net 3.60mX. 3.60m
h) Spacing	: 90 cm X 60 cm
i) No. of plants/plot	: Gross: 48 Net: 24
j) Date of sowing	: 04-12-2015
k) Date of Transplanting	: 23-12-2015
l) Manures	: 20t/ha
m) Fertiliser dose	: RDF: - 150 : 75 : 50 kg NPK/ha
	50 kg N and full dose of P and K are applied at the time of transplanting. Remaining 100 kg of N is applied in two equal split, first at one month after transplanting and second two months after transplanting
n) Irrigations	: As per requirement (After every 6 to 8 days of interval)
o) Observations recorded	: 1) Final plant stand
	2) Plant height (cm)
	3) Days to 50% flowering
	4) No. of primary branches per plant
	5) Dry weight of roots per plant (g)
	6) No. of fruits per plant
	7) Weight of fruits per plant (kg)
	8) Single fruit weight (g)
	9) Diameter of fruit(mm)
	10) Fruit yield per plot (kg)
	11) Fruit yield (q/ha)

Results and Discussion:

The trial was conducted to know the effect of Magnesium Chelate powder 6% on fruit yield and ancillary characters of tomato crop. Spraying of Magnesium Chelate powder 6% in four different doses (0.5g, 1.0 g, 1.5 g and 2.0 g per liter of water) was undertaken along with one control. the data on fruit yield and ancillary characters are presented in Table 1.

The data presented in table I revealed that the final plant stand, plant height, days to 50% flow ering. number of primary branches per plant and weight of fruit per plant did not differ significantly due to different treatments. Spraying of Magnesium Chelate powder 6%@ 1.5 g per liter of water produced significantly more dry root weight per plant over spraying of Magnesium Chelate powder 6% @ 1.0 g and 0.5 g per liter of water. Significantly higher number of fruit per plant and diameter of fruit were recorded in treatment of spraying of Magnesium Chelate powder 6% @ 0.5 g per liter of water (T2) over rest of the treatment under study. except treatment T3 and T1 for number of fruit per plant and treatment T3 for diameter of fruit which remained statistically at par with treatment T2. Significantly maximum single fruit weight, fruit yield per plot and fruit yield per hectare were recorded in treatment of spraying of Magnesium chelate powder 6%@ 1.0 g per liter of water (T3) over rest of the treatment under study, except treatment T2 for single fruit weight and treatments T2 and T4 for fruit yield per plot and fruit yield per hectare which remained statistically on par with T3.

Table1 'Yield and Ancillary data of product (Magnesium chelate powder 6%) Testing, Trial on Tomato conducted during Rabi 2015-16

Treatment Details	No.of fruits per plant	Wt. of fruit Per plant(Kg)	Single fruit wt.(g)	Diameter of fruit (mm)	Fruit yield per plot(kg)	Fruit Yield(Q/ha)
T1: Control	63	1.286	21.1	43	25.4	196
T2: Spraying of Magnesium Chelate powder 6% @ 0.5 g per liter of water	67	1.294	22.0	45	26.0	200
T3: Spraying of Magnesium Chelate powder 6%@ 1.0g per liter of water	66	1.382	23.3	44	27.8	214
T4: Spraying of Magnesium Chelate powder 6% @1.5g per liter of water	60	1.235	21.2	43	26.3	203

T5: Spraying of Magnesium Chelate Powder 6% @2.0 g per liter of water	59	1.179	20.5	42	21.6	167
Mean	63.28	1.28	21.62	43.24	25.41	196.20
SE(+)	1.41	0.05	0.61	0.63	0.69	5.36
CD at 5%	4.24	N.S	1.82	1.90	2.07	16.07

Treatment Details	Final plant stand per plot	Plant Height (cm)	Days to 50% flowering	No of Primary Branches per plant	Dry root weight per plant (g)
T1: Control	19	90	39	8	21.7
T2: Spraying of Magnesium Chelate powder 6% @ 0.5 g per liter of water	18	89	40	8	18.0
T3: Spraying of Magnesium Chelate powder 6%@ 1.0g per liter of water	19	87	39	8	18.6
T4: Spraying of Magnesium Chelate powder 6% @1.5g per liter of water	19	91	39	9	23.1
T5: Spraying of Magnesium Chelate Powder 6% @2.0 g per liter of water	19	87	40	8	22.7
Mean	18.68	89	39.48	8.12	20.18
SE(+)	0.33	1.38	0.56	0.33	1.03
CD at 5%	NS	NS	NS	NS	3.10

Conclusion

From the results of one year data, it is observed that significantly higher fruit yield of tomato was recorded when crop was sprayed with Magnesium Chelate powder 6%@ 1.0 g per liter of water.

TRIAL REPORT 14

Name of Organisation : Mahatma Phule Krishi Vidyapeeth Rahuri

Location of testing : AICRP on Vegetable Crops, Department of HorticultureM, PKV., Rahuri

Product details : Evaluation of Ferrous Chelate (6%) for growth and yield in Okra

Name of Firm : M/s Agro Inputs Manufacturer Association of (India), Pune, Maharashtra India.
ChhatrapatSi hivajiM arketY ard, Shop N o.6, 1ˢᵗ floor, Madhavarti Bhavan, Gultekadi, Pune-4Il 037.

Test Period : February2 014 to May, 2014

Details of Experiment

    a. No. of Treatments : Four
    b. Treatment details

T. No.	Treatments	Dose (g/liter)
L I.	Ferrous Chelate (6%)	0.5
	Ferrous Chelate (6%)	1.0
	Ferrous Chelate (6%)	1.5
	Control	'--

    c. Experimental design :Randomised Block Design (RBD)
    d. No. of replications :Five
    e. Plot size :7.2 m2
    f. Soil type :Medium black

Crop details

    a. Crop/Variety   Phule Utkarsha
    b. Date of sowing   28.02.2014
    d. Spacing   60 x 30 cm

e. Date of harvestings   16.04.2014 to 26.05.2014

Methodology:

a. Spraying equipment   Knapsack sprayer
b. Method of application: Foliar application

Time of Application   Three sprays
    a. At 21-30 days after sowing
    b. At 50 % flowering
    c. At fruit development stage

Observation Taken   : i) Plant ht (cm)
    ii) No. of fruits/plant
    iii) Fruit weight (g)
    iv) Fruit length (cm)
    v) Fruit diameter (cm)
    vi) Yield /plot (kg)
    vii) Yield (q/ha)
    viii) Incidence of diseases and Pest

Results

The Okra variety 'Thule Utkarsha" was used to study the effect of Ferrous Chelate (6%) sprays of different concentrations on yield during summer 2014. The observations on growth and yield parameters were given in Table 1. Among the different treatments studied, the maximum fruit yield was observed in treatment T2 (177.53 q/ha) i.e. Ferrous Chelate (6%) at 1.0 g per liter at 30 days after sowing, 50 % flowering and at fruit development stage which was at par with the treatment T3 (174.13 q/ha) i.e. Ferrous Chelate (6%) at 1.5 g per liter at 30 days after sowing, 50 % flowering and at fruit development stage. The lowest yield was observed in T4 (156.23 q/ha) i.e. Control treatment followed by T1 i.e. Ferrous Chelate (6%) at 0.5 g per liter at 30 days after sowing, 50 % flowering and at fruit development stage. All the doses of Ferrous chelate (6%) recorded the highest yield than the only control treatment.

Table1: Effect of different concentration of Feerous Chelate (6%) on growth & yield of Okra

Treat No.	Plant ht. (cm)	No of fruits/ plant	Fruit Length (cm)	Fruit Diameter (cm)	Av. Fruit wt.(g)	Fruit Colour	Yield/ Plant (gm)	Yield Kg/plot	Yield q/ha	YVMV Incidence (%)	Fruit Borer (%)
T1	67.69	24.35	12.19	1.43	12.36	Light Green	382.7	11.48	159.36	19.38	32.74
T2	65.5	23.3	12.63	1.44	12.98	Light Green	426.4	12.79	177.53	12.75	34.07
T3	69.64	26.36	12.42	1.44	12.78	Light Green	418.2	12.54	174.13	15.94	33.00
T4 (Control)	67.46	23.32	11.81	1.43	11.98	Light Green	375.2	11.25	156.23	21.68	36.12
SE+_	1.62	1.01	0.15	NS	0.11		6.63	0.19	2.76		
CD at 5%	5.01	3.12	0.47	NS	0.35		20.44	0.61	8.51		

Conclusion

In study of Ferrous Chelate (6%) sprays at different concentrations in Okra during summer 2014, the treatment T2 i.e. Ferrous Chelate (6%) at 1.0 g per liter at 30 days after sowing, 50 % flowering and at fruit development stage recorded the maximum fruit yield of (177.53 q/ha.) which was at par with the treatment T3 (174.13 q/ha) i.e. Ferrous Chelate (6%) at 1.5 g per liter at 30 days after sowing, 50 % flowering and at fruit development stage.

TRIAL REPORT 15

Name of Organisation	: Dr. Balasaheb Sawant Konkan Krishi Vidyapeeth Dapoli 415712, Dist.Ratnagiri (M.S.)
Product Details	: Ferrous chelate (6%) (Fe EDDHA)
Treatment Details and Company	: T1- Spraying of Ferrous Chelate @ 0.5 g/L T2- Spraying of Ferrous Chelate @ 1.0 g/L T3- Spraying of Ferrous Chelate @ 1.5 g/L T4-Control (Recommended Package of Practices) M/s Agro Inputs Manufacturer Association of (India), Pune, Maharashtra India.
Replications	: Five (5)
Design	: Randomized Block Design
Plot Size	: 3.0 m X 2.7 m
Observations	: 1. Plant Height (cm) 2. No. of branches 3. Length of fruit (cm) 4. Girth of fruit (cm) 5. Average wt. of fruit (kg) 6. No. of fruits per plant 7. No. of fruits per plot 8. Yield per plant (Kg) 9. Yield per plot (kg) 10. Yield per hectare (t)

Methodology:

The spraying of Ferrous Chelate was done at 20 days, 40 days and 60 days after sowing as per the treatment details. All the package of practices was followed as per the university recommendations for Okra. The observations were recorded on ten randomly selected plants per treatment per replications and the average of this was worked out. The data recorded were analysed by the methods suggested by Panse and Sukhateme (1995).

Results

The results indicated that the height and number of branches was not influenced significantly due to spraying of ferrous chelate at various -concentrations. The number of fruits per plant affected significantly and maximum number of fruits per plant (25.26) and per plot (842.60) were recorded maximum in treatment T3 - Spraying of ferrous Chelate @ 1.5 g /L. The yield per plant and yield per plot was affected significantly with due to spraying of ferrous chelate at various concentrations. The spraying of ferrous Chelate @ 1.5 g /L (T3) recorded maximum yield, per plant (0.348 kg), per plot (12.53 kg) and per hectare (15.47 t).

Table No.1 Effect of Ferrous chelate (6%) (Fe EDDHA) on growth of okra

Treatments	Height of plant (cm)	Number of branches	Length of Fruit(cm)	Girth of Fruit (cm)	Avg. Fruit Weight(Kg)
T1	108.56	1.42	15.20	1.80	0.020
T2	109.92	1.38	16.10	1.84	0.020
T3	109.58	1.48	16.66	1.81	0.019
T4	106.58	1.28	16.84	1.86	0.020
SE+	2.61	0.07	0.38	0.01	0.09
CD at 5 %	NS	NS	1.17	0.03	NS

Table No 2: Effect of ferrous chelate (6%) (Fe EDDHA) on yield attributing characters of Okra

Treatments	No of Fruits per plant	No. of Fruits per plot	Yield per plant(kg)	Yield per plot(kg)	Yield per hectare(t)
T1	20.03	677.80	0.284	10.24	12.64
T2	24.40	797.20	0.331	11.97	14.79
T3	25.26	842.60	0.348	12.53	15.47
T4	21.40	737.00	0.298	10.71	13.23
SE+	1.21	46.56	0.02	0.63	0.66
CD at 5 %	3.73	143.47	0.07	1.85	20.03

TRIAL REPORT 16

Name of Organisation	: National Horticultural Research and Development Foundation, Nashik (MS)
Name of the Company	: Aries Agro Limited
Product Details	: Multi micronutrient Formulation
Crop and Variety	: Tomato, Arka Rakshak (F-1 Hybrid)
Season	: Kharif -2013
Experimental Design	: Randomized Block Design
Replication	: Three (03)

MATERIALS AND METHODS

The experiment was conducted on tomato crop var. Arka Rakshak (F1) during the kharif 2013 at RRS, Chitegaon, Nashik. The experiment was laid out in Randomized Block Design with three replications. The soil nutrient status of the experimental site was tested 86 given below:

PH - 7.62   EC   0.141dSm^{-1}
Organic Carbon -   0.50 %   Av. Nitrogen   -   234 kg/ha
Av. Phosphorus -   49.05 kg/ha   Av. Potash   -   336 kg/ha
Calcium Carbonate- 3.7 %

The treatments were comprised of

Ti -   (Application of Multi micronutrient Formulation at 21 DAT through drenching @ 250 ml/acre.),

T2-   (Application of Multi micronutrient Formulation at 21 and 45 DAT through drenching @ 250 ml/acre.),

T3-   (Application of Multi micronutrient Formulation at 21 DAT through drenching @ 250 ml/acre and foliar application @ 1 ml/lit.),

T4 - (Application of Multi micronutrient Formulation at 45 DAT through drenching @ 250 ml/acre and foliar application @ 1 ml/lit.),

T5- Application of Multi micronutrient Formulation at 21 86 45 DAT through drenching @ 250 ml/acre and foliar application @ 1 ml/lit.),

T6- (Foliar application of Multi micronutrient Formulation at 21 DAT @ 1 ml/lit.),

T7- (Foliar application of Multi micronutrient Formulation at 45 DAT @ 1 ml/lit.),

T8 - (Foliar application of Multi micronutrient Formulation at 21 8s 45 DAT @ 1 ml/lit.) and

T9- (Control- no Multi micronutrient Formulation). The data were recorded on growth, yield and quality parameters 86 are presented in Table 1.

Result

The data presented in Table 1 revealed that all the recorded characters showed non-significant variations due to different treatments except no of fruit per plant, average weight of fruit, yield 86 marketable yield. Significantly maximum number of fruit per plant (60 nos.) were recorded in the treatment T7 (Foliar application of Multi micronutrient Formulation at 21 DAT @ lml/lit) 86 found at par with the treatments 13 (Application of Multi micronutrient Formulation at 21 DAT through drenching @ 250 ml /acre 85 foliar application @ 1 ml/lit.) 85 T9 (Control- no Multi micronutrient Formulation). Maximum average weight of fruit (51 gm) was recorded in the treatment 16 86 found at par with the treatments Ti, T2, T3, T4, T5 85 T8. Maximum gross yield (576.7 q/ha) 86 marketable yield (533.3 q/ha) were recorded in the treatment T3 85 found at par with the treatments T4 (Application of Multi micronutrient Formulation at 45 DAT through drenching @ 250 ml/acre 86 foliar application @ 1 ml/lit.) 86 T5 (Application of Multi micronutrient Formulation at 21 86 45 DAT through drenching @ 250 ml/acre 86 foliar application @ 1 ml/lit.) in respect of gross yield only.

Table: 1: Evaluation of Multi micronutrient Formulation against growth & yield of Tomato crop

Treatment	Height of plant at 60 DSP	Days for flowering	Days for first harvest	No. of fruits per plant (no)	Diameter of fruit	Total No of pickings	Av. wt of fruit (gm)	Gross yield (q/ha)	Marketable yield (q/ha)	Shelf life at ambient room temp. (days)	TSS (%)	Phyto toxicity
T 1	112.2	38.7	70	53	4.7	17.0	47	501.7	463.0	7	4	Nil
T 2	116.9	38.7	70	56	4.6	17.0	46	515.6	466.0	7	4	Nil
T 3	112.5	38.0	70	59	4.6	17.0	49	576.7	533.3	7	4	Nil
T 4	115.5	38.7	71	56	4.7	16.7	49	542.2	488.5	7	4	Nil
T 5	112.5	39.3	71	55	4.7	16.7	50	544.4	496.5	7	4	Nil
T 6	116.1	39.3	71	50	4.6	16.7	51	515.2	458.1	7	4	Nil
T 7	116.0	39.3	70	60	4.6	17.0	40	487.0	431.8	7	4	Nil
T 8	115.6	39.3	71	51	4.7	16.7	46	466.5	413.4	7	4	Nil
T 9	111.9	38.0	71	57	4.8	16.7	43	497.1	448.4	7	4	Nil

S.E.m+	3.7	0.87	0.9	1.8	0.1	0.3	2.6	17.4	16.5	0.3	0.1	Nil
C.D at 5%	NS	NS	NS	3.8	NS	NS	5.5	36.8	35.0	NS	NS	Nil
C.V (%)	3.9	2.76	1.6	3.9	3.4	2.3	6.8	4.1	4.3	4.8	2.4	Nil

T1 - Application of Multi micronutrient Formulation at 21 DAT through drenching @ 250 ml/acre

T2 - Application of Multi micronutrient Formulation at 21 & 45 DAT through drenching @ 250 ml/acre

T3 - Application of Multi micronutrient Formulation at 21 DAT through drenching @ 250 ml/acre & foliar application @ 1 ml/lit.

T4 - Application of Multi micronutrient Formulation

T5 - Application of Aries total at 21 & 45 DAT through drenching @ 250 ml/acre & foliar application @ 1ml/lit.

T6 - Foliar application of Multi micronutrient Formulation at 21 DAT @ 1 ml/lit.

T7 - Foliar application of Multi micronutrient Formulation at 45 DAT @ 1 ml/lit.

T8 - Foliar application of Multi micronutrient Formulation at 21 & 45 DAT @ 1 ml/lit.

T9 - Control (no Multi micronutrient Formulation)

CONCLUSION

The study conducted on tomato variety Arka Rakshak at RRS, Nashik during khaif-2013 revealed that application of Multi micronutrient Formulation at 21 days after transplanting through drenching @ 250 ml/acre and foliar application @ lml/lit enhance the fruit yield of tomato.

TRIAL REPORT 17

Name of Organisation : National Horticultural Research and Development Foundation, Nashik (MS)

Name of the Company : Aries Agro Limited

Product Details : Multi micronutrient Formulation containing N,P,K,Cu,Mn,B and Mo

Crop and Variety : Tomato, Arka Rakshak (F-1 Hybrid)

Season : Kharif -2014

Experimental Design : Randomized Block Design

Replication : Three (03)

Date of Transplanting : 30/8/2014

Date of First Flowering : 7/10/2014

Date of First Fruit Setting : 10/10/2014

The soil nutrient status of the experimental site is given below:

pH	7.47
EC	0.446 dSm^{-1}
Organic Carbon	0.50 %
Available Nitrogen	234 kg/ha
Available Phosphorus	58.86 kg/ha
Available Potash	336 kg/ha
Calcium Carbonate	5.0 %
Available Calcium	800 ppm
Magnesium	288 ppm
Available Sodium	200 ppm

Sulphur	20.93 mg/kg
Copper	1.87 mg/kg
Zinc	0.39 mg/kg
Iron	12.4 mg/kg
Manganese	18.9 mg/kg

The treatments were comprised of

T1 - (Application of Multi micronutrient Formulation at 21 DAT through drenching @ 250 ml/acre.),

T2 - (Application of Multi micronutrient Formulation at 21 & 45 DAT through drenching @ 250 ml/acre.),

T3 - (Application of Multi micronutrient Formulation at 21 DAT through drenching @ 250 ml/acre and foliar application @ 1 ml/lit.),

T4 - (Application of Multi micronutrient Formulation at 45 DAT through drenching @ 250 ml/acre and foliar application @ 1 ml/lit.),

T5 - (Application of Multi micronutrient Formulation at 21 & 45 DAT through drenching @ 250 ml/acre and foliar application @ 1 ml/lit.),

T6 - (Foliar application of Multi micronutrient Formulation at 21 DAT @ 1 ml/lit.),

T7 - (Foliar application of Multi micronutrient Formulation at 45 DAT @ 1 ml/lit.),

T8 - (Foliar application of Multi micronutrient Formulation at 21 and 45 DAT @ 1 ml/lit.) and

T9 - (Control- no Multi micronutrient Formulation)

The data were recorded on growth, yield & quality parameters and are presented in Table 1.

Table 1: Evaluation Of Aries Total Against Growth And Yield Of Tomato Crop

Treatment	Height of plant at 60 DAT	Days of first flower initiation	Days of first harvest	Total No of pickings (No)	Diameter of fruit (cm)	Length of fruit (cm)	No of fruit per plant (no)
T1	88.6	39	72	12	4.1	4.9	35.2
T2	83.6	39	73	12	4.3	4.9	30.8
T3	86.1	38	73	12	4.3	4.8	40.2
T4	83.6	39	73	12	4.0	4.8	34.2
T5	87.9	39	73	12	4.4	4.9	49.6
T6	86.1	39	72	12	4.2	4.8	30.9
T7	81.6	39	73	12	4.3	5.0	34.0
T8	86.9	39	72	12	4.3	4.9	31.3
T9	80.9	38	73	12	4.1	4.7	30.2
S.E.m+	6.0	0.8	0.6	0.3	0.2	0.2	2.4
C.D at 5%	NS	NS	NS	NS	NS	NS	5.1

Treatment	Av. Wt. of fruit (gm) (10 no of fruits)	Gross yield (q/ha)	Marketable yield (q/ha)	Shelf life of fruit (days)	TSS (%)	Phyto toxicity at 1, 3, 5, 7 & 10 days after application of treatment
T1	69.8	363.9	348.9	7	4.0	NIL
T2	73.3	347.4	330.8	7	4.0	NIL
T3	77.3	377.0	358.6	7	4.0	NIL
T4	73.6	376.4	355.1	7	4.0	NIL
T5	78.3	411.3	395.6	7	4.0	NIL
T6	77.6	337.9	323.8	7	4.0	NIL
T7	65.7	326.2	306.9	7	4.0	NIL
T8	69.6	366.5	348.4	7	4.0	NIL
T9	53.2	319.0	298.7	7	4.0	NIL
S.E.m+	5.2	18.9	21.2	0.3	0.0	NIL
C.D at 5%	11.0	40.2	45.0	NS	NS	NIL

Treatments:

T1 - (Application of Multi micronutrient Formulation at 21 DAT through drenching @ 250 ml/acre.),

T2 - (Application of Multi micronutrient Formulation at 21 and 45 DAT through drenching @ 250 ml/acre.),

T3 - (Application of Multi micronutrient Formulation at 21 DAT through drenching @ 250 ml/acre and foliar application @ 1 ml/lit.),

T4 - (Application of Multi micronutrient Formulation at 45 DAT through drenching @ 250 ml/acre and foliar application @ 1 ml/lit.),

T5 - (Application of Multi micronutrient Formulation at 21 & 45 DAT through drenching @ 250 ml/acre and foliar application @ 1 ml/lit.),

T6 - (Foliar application of Multi micronutrient Formulation at 21 DAT @ 1 ml/lit.),

T7 - (Foliar application of Multi micronutrient Formulation at 45 DAT @ 1 ml/lit.),

Ts - (Foliar application of Multi micronutrient Formulation at 21 and 45 DAT @ 1 ml/lit.) and

T9 - (Control- no Multi micronutrient Formulation)

RESULT

The data presented in Table 1 revealed that all the recorded characters showed non- significant variation between except number of fruit per plant, average weight of fruit, gross yield & marketable yield. Significantly highest number of fruit per plant (49.6), average weight of 10 fruit (78.3 gm), gross yield (411.3 q/ha) & marketable yield (395.6 q/ha) were recorded in the treatment T5 (Application of Multi micronutrient Formulation at 21 & 45 DAT through drenching @ 250 ml/acre & foliar application @ 1 ml/lit.) & found at par in respect of average fruit weight with the treatments T6 (Foliar application of Multi micronutrient Formulation at 21 DAT @ 1 ml/lit.), T3 (Application of Multi micronutrient Formulation at 21 DAT through drenching @ 250 ml/acre & foliar application @ 1 ml/lit), T4 (Application of Multi micronutrient Formulation at 45 DAT through drenching @ 250 ml/acre & foliar application @ 1 ml/lit), T2 (Application of Multi micronutrient Formulation at 21 & 45 DAT through drenching @ 250 ml/acre), Ti (Application of Multi micronutrient Formulation at 21 DAT through drenching @ 250 ml/acre), & T8 (Foliar application of Multi micronutrient Formulation at 21 & 45 DAT @ 1 ml/lit), whereas in respect of gross & marketable yield with the treatments T3 (377.0 & 358.6 q/ha) & T4 (376.4 &355.1 q/ha).

CONCLUSION

The study conducted on evaluation of of Multi micronutrient Formulation against growth & yield of tomato variety Arka Rakshak (F-1 hybrid) at RRS, Nashik during Kharif, 2014 revealed that application of of Multi micronutrient Formulation at 21 & 45 days after transplanting through drenching @ 250 ml/acre & foliar application @ 1 ml/lit increases the yield of tomato by 29 % as compared to control.

TRIAL REPORT 18

Name of Organisation : National Horticultural Research and Development Foundation, Nashik (MS)

Name of the Company : Aries Agro Limited

Product Details : Liquid Chelated Micronutrient Mixture

Crop and Variety : Okra, Arka Anamika

Season : Kharif -2013

Experimental Design : Randomized Block Design

Replication : Three (03)

Date of Sowing : 01/08/2013

Total Pickings : 22

The soil nutrient status of the experimental site is given below:

pH	7.75
EC	0.13 dSm^{-1}
Organic Carbon	0.60 %
Available Nitrogen	290 kg/ha
Available Phosphorus	68.67 kg/ha
Available Potash	336 kg/ha
Calcium Carbonate	3.6 %

The treatments were comprised of

T1 - (Application of Liquid Chelated Micronutrient Mixture at 20 DAS@ 500 ml/acre.),

T2 - (Application of Liquid Chelated Micronutrient Mixture through drip at 45 DAS@ 500 ml/acre.),

T3 - (Foliar Application of Liquid Chelated Micronutrient Mixture at 20 DAS @ 1 ml/lit.),

T4 - Foliar Application of Liquid Chelated Micronutrient Mixture at 45 DAS @ 1 ml/lit.),

T5 - (Application of Liquid Chelated Micronutrient Mixture at 20 DAS through drip @ 500 ml/acre and foliar application @ 1 ml/lit.),

T6 - (Application of Liquid Chelated Micronutrient Mixture at 45 DAS through drip @ 500 ml/acre and foliar application @ 1 ml/lit.),

T7 - (Application of Liquid Chelated Micronutrient Mixture through drip at 20 and 45 DAS @ 500 ml/acre),

T8 - (Foliar application of Liquid Chelated Micronutrient Mixture at 20 and 45 DAS @ 1 ml/lit.) and

T9 - (Control- no Liquid Chelated Micronutrient Mixture)

The data were recorded on growth, yield & quality parameters and are presented in Table 1

RESULT

The data presented in Table 1 revealed that all the characters exhibit non-significant vaiations due to different treatments except height of plant at 60 days after sowing, gross yield and marketable yield. Maximum height of plant (121 cm) was recorded in the treatment T1 - (Application of Liquid Chelated Micronutrient Mixture at 20 DAS@ 500 ml/acre.) and found at par with the treatment T3 (Foliar Application of Liquid Chelated Micronutrient Mixture at 20 DAS @ 1 ml/lit.). Significantly maximum gross yield (192.6 q/ha) and marketable yield (157.5 q/ha) were recorded in the treatment T6 (Application of Liquid Chelated Micronutrient Mixture at 45 DAS through drip @ 500 ml/acre and foliar application @ 1 ml/lit.), the marketable yield was found at par with T3.

Table 1: Evaluation of Liquid Chelated Micronutrient Mixture against growth and yield of Okra

Treatment	Height of plant at 60 DAS	Days for flowering	Days for first harvest	Length of fruit (cm)	Diameter of fruit (cm)	No of pickings	Gross yield (q/ha)	Marketable yield (q/ha)	Shelf life at room temp	Phytotoxicity
T1	121.0	42.7	47	9.0	1.4	22.0	179.8	145.6	4	NIL
T2	113.7	42.0	47	8.8	1.5	22.0	181.7	147.4	5	NIL
T3	116.4	42.3	47	9.6	1.4	22.0	183.3	149.8	5	NIL
T4	106.1	42.7	48	9.2	1.4	22.0	168.2	136.6	5	NIL
T5	114.0	42.0	47	9.4	1.5	22.0	176.5	142.8	4	NIL
T6	106.1	42.7	47	8.9	1.4	22.0	192.6	157.5	5	NIL
T7	113.7	42.3	48	9.5	1.3	22.0	163.3	132.8	4	NIL
T8	104.5	42.7	47	9.9	1.5	22.0	174.0	140.3	4	NIL
T9	106.6	42.7	47	8.7	1.3	22.0	179.2	148.0	4	NIL
S.E.m±	3.1	0.4	0.6	0.6	0.1	0.2	4.3	4.2	0.4	NIL
C.D at 5%	6.6	NS	NS	NS	NS	NS	9.2	9.0	NS	NIL
C.V (%)	3.4	1.2	1.6	8.5	6.6	1.2	3.0	3.6	11.6	NIL

T1 - (Application of Liquid Chelated Micronutrient Mixture at 20 DAS@ 500 ml/acre.),

T2 - (Application of Liquid Chelated Micronutrient Mixture through drip at 45 DAS@ 500 ml/acre.),

T3 - (Foliar Application of Liquid Chelated Micronutrient Mixture at 20 DAS @ 1 ml/lit.),

T4 – (Foliar Application of Liquid Chelated Micronutrient Mixture at 45 DAS @ 1 ml/lit.),

T5 - (Application of Liquid Chelated Micronutrient Mixture at 20 DAS through drip @ 500 ml/acre and foliar application @ 1 ml/lit.),

T6 - (Application of Liquid Chelated Micronutrient Mixture at 45 DAS through drip @ 500 ml/acre and foliar application @ 1 ml/lit.),

T7 - (Application of Liquid Chelated Micronutrient Mixture through drip at 20 and 45 DAS @ 500 ml/acre),

T8 - (Foliar application of Liquid Chelated Micronutrient Mixture at 20 and 45 DAS @ 1 ml/lit.) and

T9 - (Control- no Liquid Chelated Micronutrient Mixture)

CONCLUSION:

The study conducted on Okra variety Arka Anamika at RRS, Nashik during Kharif-2013 revealed that application of Liquid Chelated Micronutrient Mixture at 45 days after sowing through drip @ 500ml/acre and foliar application @1ml/lit enhance yield and quality of Okra fruits, however, marketable yield can be obtained at par by foliar application of Liquid Chelated Micronutrient Mixture at 20 DAS@ 1ml/lit.

TRIAL REPORT 19

Name of Organisation : National Horticultural Research and Development Foundation, Nashik (MS)

Name of the Company : Aries Agro Limited

Product Details : Effect of Amino acid chelated Zinc and Zinc-EDTA against Zinc Sulphate

Crop and Variety : Tomato, Tt0DHYB-6

Season : Kharif-2012

Experimental Design : Randomized Block Design

Bed Size : 7.5 m x 1.2 m (raised beds)

Date of Transplantation : 30/07/2012

Date of flowering : 18/08/2012

Date of first picking : 27/09/2012

Date of last picking : 26/10/2012

Treatments

T1- Basal application of amino acid chelated Zinc @ 500 g/acre
'T2- Basal application of Zinc-EDTA @ 500 qm/acre •
T3- Basal application of zinc sulphate @ 10 kg/acre
T4- Application of amino acid chelated Zinc through drip and foliar application at 45 DAT @ 500 g/acre through drip arid 2 g/lit. of water for spray.
T5- Application of Zinc-EDTA through drip and foliar application at 45 DAT @ 500 g/acre through drip and 1 g/lit of water for spray.
T6- Application of zinc sulphate through drip and foliar application at 45 DAT @ 10 kg/acre through drip and 3 g/lit. of water for spray.

T7- Application of amino acid chelated Zinc through drip at 21 DAT @ 500 g/acre.
T8- Application of Zinc-EDTA through drip at.21 DAT @ 500 g/acre.
T9- Application of zinc sulphate through drip at 21 DAT @ 10 kg/acre.
T10- Foliar application of amino acid chelated Zinc at 21 & 45 DAT 2 g/lt. of water.
T11- Foliar application of Zinc-EDTA at 21 & 45 DAT 1g/lt. of water.
T12- Foliar application of zinc Sulphate at 21 & 45 DAT 3 g/lit. of water.
T13- Control (No Zn)

RESULT AND DISCUSSION

The flowering had started on 18/08/2012 and first picking of fruits was done on 27/09/2012 while last picking was done on 26/10/2012. During the period of about one months total 7 pickings were made. The data presented in Table-1 revealed that although gross yield did not show significant differences the significantly highest marketable yield (594.92 q/ha) was recorded in T2 (basal application of Zinc- EDTA @ 500 g/ac.) indicating Zinc-EDTA's effect on quality improvement in tomato. The plant stand at flowering as well as one month showed non-significant variations due to various treatments. The plant height at 30 DAP (99.80 cm) as well as 60 DAP (123.35 cm) were recorded significantly highest in T6 (application of zinc sulphate through drip and foliar application at 45 DAP @ 10 kg/acre through drip and 3 g/lit of water for spray) which were found at par with T7 (application of amino acid chelated Zinc through drip at 21 DAP @ 500 g/oc.) at 30 DAP and T3 (basal application of zinc sulphate @ 10 kg/ac.) at 60 DAP. All the treatments except T9 (application of $ZnSO_4$ through drip at 21 DAP @ 10 kg/acre) showed significant increase in plant height over control indicating treatment effect on plant growth. However, no significant variations were recorded for number of fruits/plant, fruit diameter and TSS contents due to different treatments. Significantly highest weight of fruits (243.65 g/5 fruits)) was recorded in T9 but it had less number of fruits and found to be at par with T1 (basal application of amino acid chelated Zinc @ 500 g/acre), T2, T4 (application of amino acid chelated Zinc through drip and foliar application at 45 DAP @ 500 g/ac. through drip and 2 gm/lit of water for spray), T5 (application of Zinc-EDTA through drip and foliar application at 45 DAP @ 500 g/ac. through drip and 1 g/lit of water for spray), T7 and T11 (foliar application of Zinc-EDTA at 21 and 45 DAP @ 1 g/lit of water). The acid-sugar ratio was found significantly highest (12.03) in T8 (application of Zinc-EDTA through drip @ 21 and 45 DAP @ 500 g/ac.) and found to be at par with T2, T3, T9 as well as T13 (control — no Zn). Lowest acid sugar ratio (7.35) was recorded for T4 (application of amino acid chelated Zinc through drip and foliar application at 45 DAT @ 500 ml/acre and 2 g/lit of water for spray. Further the storage of the produce was found good till 14 days under ambient conditions, thereafter deterioration started interms of shrinkage, loosening and roting.

Among the basal application of amino acid chelated Zinc (T1), Zinc-EDTA (T2) and zinc sulphate (T3), the maximum marketable yield (594.92 q/ha) and average weight of 5 fruits (221.80 gm) were recorded in T2 while acid-sugar ratio (11.41) was highest in T3 indicating Zinc-EDTA as better for soil application.

Similarly in treatment of amino acid chelated Zinc (T4), Zinc-EDTA (T5) and zinc sulphate (T6) through drip and foliar application, the maximum marketable yield (544.97 q/ha) and acid sugar ratio (10.00) were recorded in T6 while average weight of 5 fruit (232.82 g) in T4 indicating Zinc Sulphate better when given through drip and foliar application.

Among the treatment of amino acid chelated Zinc (T7), Zinc-EDTA (T8) and zinc sulphate (T9) through drip, the maximum marketable yield (538.26 q/ha) was recorded in T7 while average weight of 5 fruits in T9 (243.65 g) and acid sugar ratio in T8 (12.03) indicating better effect of amino acid chelated Zinc when applied through drip.

In the treatment of foliar application at 21 and 45 DAP with amino acid chelated Zinc (T10), Zinc-EDTA (T11) and zinc sulphate (T12), the maximum marketable yield (503.34 q/ha) and acid sugar ratio (10.25) were recorded in T10 while T11 recorded the maximum average weigh of 5 fruits (215.49 gm) indicating amino acid chelated Zinc as better when applied through foliar application.

Effect of Pro zinc and chelamin agaist zinc sulphate on growth, yield and quality of tomato at RRS Nashik during Kharif 2012

Treatments	Gross Yield q/h8	ñ4arkcteble yield q/ha	Pleat staod at flowering	Plaat heibent (cm) at 30 DAT	PTaot height (czn) at6O DAT	Number of Orbits/plgnt0	A>erege aeightoF 5 Fruits(gm)	Fruit diameter (cm)	TSS %	Acid sugar ratio
T1	554.36	527.91	30.00	93.80	114.60	18.05	203:46	4.52	3.73	9.97
T2	6T4.80	594.92	29.00	'94.85	1|7.70	19.44	221.80	4.92	3.57	10.74
T3	571.06	558.55	29.00	96.40	121.05	17.21	168.95	4.06	3.80	11.43
T4	522.91	498.50	30.00	95.60	118.40	16.4	232.82	S.19	3.34	7.35

T5	517.14	504.18	29.10	89.20	115.70	16.27	202.20	4.64	3.34	8.99
T6	557.81	544.97	29.00	99.80	123.35	18.41	164.84	4.05	3.90	10.00
T7	552.39	538.26	29.50	97.80	120.90	17.12	209.92	4.75	3.70	10.82
T8	489.48	469.78	29.00	94.80	115.85	15.69	184.82	4.35	3.97	12.03
T9	485.84	472.8.1	29.50	83.85	112.60	15.1	243.65	5.36	3.70	11.25
410	519.22	504.34	29.50	96.40	113.05	13.84	194.S6	4.39	3.77	10.2S
TI1	473..57	454.01	28.50	92.65	111.90	15.37	215.49	4.78	3.50	8.55
T12	516.52	487:38	30.00	93.85	114.95	17.23	139.89	4.04	3.20	9.64
T13	578.18	559.33	29.00	84.80	101.45	17.76	188.99	4.22	3.37	10.88
S.Em*	57.64	4.53	0.73	1.44	1.07	1.59	21.42	0.44	0.35	0.66
CDut5%	NS	9.87	NS	3.14	2.33	NS	46.67	NS	t4S	1.44
CV%	119.75	0.88	2.5	1.54	0.92	9.48	10.83	9.65	9.62	6.S3

Treatments

T1- Basal application of amino acid chelated Zinc @ 500 g/acre
'T2- Basal application of Zinc-EDTA @ 500 qm/acre
T3- Basal application of zinc sulphate @ 10 kg/acre
T4- Application of amino acid chelated Zinc through drip and foliar application at 45 DAT @ 500 g/acre through drip arid 2 g/lit. of water for spray.
T5- Application of Zinc-EDTA through drip and foliar application at 45 DAT @ 500 g/acre through drip and 1 g/lit of water for spray.
T6- Application of zinc sulphate through drip and foliar application at 45 DAT @ 10 kg/acre through drip and 3 g/lit. of water for spray.
T7- Application of amino acid chelated Zinc through drip at 21 DAT @ 500 g/acre.
T8- Application of Zinc-EDTA through drip at.21 DAT @ 500 g/acre.
T9- Application of zinc sulphate through drip at 21 DAT @ 10 kg/acre.
T10- Foliar application of amino acid chelated Zinc at 21 & 45 DAT 2 g/lt. of water.
T11- Foliar application of Zinc-EDTA at 21 & 45 DAT 1g/lt. of water.
T12- Foliar application of zinc Sulphate at 21 & 45 DAT 3 g/lit. of water
T13- Control (No Zn)

CONCLUSION

The study conducted at RRS, Nashik during kharif,2012 on tomato hybrid TODHYB-6 revealed that amino acid chelated Zinc and Zinc-EDTA are effective in enhancing the yield and other yield parameters however, its beneficial effect depend on the status of zinc in soil and through thus their application should be based on soil status.

TRIAL REPORT 20

Name of Organisation	: National Horticultural Research and Development Foundation, Nashik (MS)
Name of the Company	: Aries Agro Limited
Product Details	: Amino acid chelated Micronutrient Mixture – Grade 2
Crop and Variety	: Tomato, TOHYD-12
Season	: Kharif -2012
Experimental Design	: Randomized Block Design
Bed Size	: 7.5 m x 1.2 m (raised beds)
Date of Transplantation	: 26.07.2012
Date of flowering	: 14.08.2012
Date of first picking	: 26.09.2013
Date of last picking	: 11.10.2012

Treatments

- T1 Basal application of Amino acid chelated Micronutrient Mixture @ 500 g/acre
- T2 Application of Amino acid chelated Micronutrient Mixture @ (500 g/acre through drenching at 20 DAT
- T3 Application of Amino acid chelated Micronutrient Mixture @ 500 g/acre through drenching at 45 DAT
- T4 Foliar application of Amino acid chelated Micronutrient Mixture © 2 9/lit of water 20 DAT
- T5 Foliar application of Amino acid chelated Micronutrient Mixture @ 2 g/lit of water at 45 DAT

T6   Application of Amino acid chelated Micronutrient Mixture @ 500 g/acre through drenching and its foliar application @ 2 g/lit of water at 20 DAT

T7   Application of Amino acid chelated Micronutrient Mixture @ 500 g/acre through drenching and its foliar application @ 2 g/lit of water at 45 DAT

T8   Control (no foliar and no drenching)

RESULT

The flowering had started on 14.08.2012 and first picking of fruits was done on 26.09.2013 while last picking was done on 11.10.2012, during the period at around one and half months a total of six pickings were made.

The data presented in Table-1 revealed that significantly highest total yield (283.20 q/ha) was recorded in T6 (Application of Amino acid chelated Micronutrient Mixture @ 500 g/acre through drenching and its foliar application 2 g/lit. of water at 20 DAT). Significantly highest plant height (50.73 cm) plant canopy (2270.13 cm2) at 30 DAT, average weight of 10 fruits (0.90 kg), fruit diameter (54.82 cm) and total no. fruits (12.05) per plant were also recorded in T6. The data further revealed that significantly highest plant height (90 cm) at 60 DAT was recorded in T1 (Basal application of Amino acid chelated Micronutrient Mixture @ 500 g/acre) which was found at par with T3 (application of Amino acid chelated Micronutrient Mixture 500 g/acre through drenching at 45 DAT, T4 (foliar application of Agripro @ 2 g/lit of water at 20 DAT and T5 (foliar application of Amino acid chelated Micronutrient Mixture @ 2 g/lit of water at 45 DAT) while plant canopy was highest in T5 which was found at par with T4 and T6. For TSS content, T5 showed significantly highest value (3.17%) which was found at par with all treatments except T7 (application of Amino acid chelated Micronutrient Mixture @ 500 g/acre through drenching and its foliar application @ 2 g/lit. of water at 45 DAT) and T8 (control) while total no. of fruits/plant at par with all treatments except T1 & T8. The acid sugar ratio was recorded significantly highest (10.19) in T3 which was found at par with T2 (application of Amino acid chelated Micronutrient Mixture @ 500 g/acre through drenching at 20 DAT), T3, T4, T5 as well as T7. Further the storage of the fruits was found good for 10 days at ambient condition, thereafter deterioration started in terms of shrinkage, loosening and rotting.

Table-1. Evaluation of Amino acid chelated micronutrient mixture grade-2 on growth and yield of Tomato at RRS Nashik during Kharif 2012

Treatments	Plant height (cm) at 30 DAT	Plant canopy (cm^2) at	Plant height (cm) at 60 DAT	Plant canopy (cmA2) at	Average weight of 10 fruits (kg)	Average diameter (mm) of fruits	TSS (%) fully ripe fruits	Total number of fruits/ plant	Acid sugar ratio	Total yield (q/ha)
Ti	50.40	2077.00	90.00	3830.00	0.83	51.63	2.90	9.73	8.94	228.17
T2	46.80	2060.53	86.20	3834.00	0.85	53.31	2.90	10.19	7.47	233.96
T3	46.33	1905.80	87.60	3922.67	0.84	53.59	2.93	10.89	ʰ 10 9	243.13
T4	47.27	1961.60	87.20	3944.87	0.89	51.15	2.97	11.11	7.82	245.89
T5	49.60	2200.73	87.00	4032.33	0.87	51.07	3.17	10.11	9.11	245.59
T6	50.73	2270.13	84.53	4015.73	0.90	54.82	3.03	12.05	9.24	283.20
T7	49.53	1902.67	85.53	3643.33	0.90	53.73	2.70	11.76	8.86	274.43
T8	47.83	1589.80	81.73	3194.20	0.82	49.14	2.60	8.61	8.46	200.10
S.Em ±	1.29	30.94	1.66	50.88	0.03	1.53	0.15	0.91	0.64	2.49
CD at 5%	2.77	66.36	3.56	109.13	0.06	3.28	0.32	1.95	1.37	5.34
CV %	3.24	1.90	2.35	1.64	3.83	3.57	6.22	10.57	8.98	1.25

Treatments

T1   Basal application of Amino acid chelated Micronutrient Mixture @ 500 g/acre

T2   Application of Amino acid chelated Micronutrient Mixture @ (500 g/acre through drenching at 20 DAT

T3   Application of Amino acid chelated Micronutrient Mixture @ 500 g/acre through drenching at 45 DAT

T4   Foliar application of Amino acid chelated Micronutrient Mixture © 2 9/lit of water 20 DAT

T5   Foliar application of Amino acid chelated Micronutrient Mixture @ 2 g/lit of water at 45 DAT

T6   Application of Amino acid chelated Micronutrient Mixture @ 500 g/acre through drenching and its foliar application @ 2 g/lit of water at 20 DAT

T7   Application of Amino acid chelated Micronutrient Mixture @ 500 g/acre through drenching and its foliar application @ 2 g/lit of water at 45 DAT

T8   Control (no foliar and no drenching)

Conclusion:

The field trial conducted at Research farm, NHRDF, Chitegaon Phata, Nashik on tomato crop variety TODHYB-12 during Kharif season, 2012 revealed that the application of Amino acid chelated Micronutrient Mixture @ 500 g/acre through drenching and its foliar application © 2 g/lit. of water at 20 DAT enhanced the plant growth, fruit size, fruit diameter, number of fruits per plant and yield over control (E9) by 41.53%.

TRIAL REPORT 21

Name of Organisation : ICAR-IIHR, Hesaraghatta Lake post, Bengaluru, 560 089

Name of the Company : M/s Karnataka Agro Chemicals

Product Details : Yield Enhancers
    P1- Enriched Organic Manure
    P2 – Consortium of all essential plant nutrients
    P3- Formulation of 17 different essential amino acids
    P4- Consortium of essential plant micronutrients
    P5- Mixture of amino acids, humic acid, sea weed extract and traces of micronutrients
    P6- Mixture of amino acids, humic acid, sea weed extract and unidentified gibberellins
    P7- Plant growth regulator

Crop and Variety : Chilli hybrid Arka Meghana

Treatment details

T. No	Treatments	Time of Application
T1	P1 @ 150 kg/acre	One application prior to transplanting
T2	Control (with RDF only)	Basal as well as top dressing
T3	P3 @ 3ml/lit	Two sprays @25th and 40th days after transplanting (DAT)
T4	P4 @ 1g/lit	Two sprays @25th and 40th DAT
T5	P2 @ 2ml/lit	Two sprays @25th and 40th DAT
T6	P6 @ 1.5ml/lit	Two sprays @25th and 40th DAT
T7	P5 @ 3ml/lit	Two sprays @25th and 40th DAT
T8	P7 @ 2ml/lit	Two sprays @25th and 40th DAT

Initial Experimental Soil properties

Results

An experiment was conducted at ICAR-Indian Institute of Horticultural Research (IIHR) to study the effect of yield enhancers on growth, biomass and yield of chilli hybrid Arka Meghana. The results of the experiment are presented as below.

Growth Parameters

Data pertaining to growth parameters viz.. plant height, girth and number of branches per plant as influenced by multiplex yield enhancers in chilli hybrid Arka Meghana are recorded and presented in table 1 & 2.

Plant height and Girth

Plant height is an indicator of growth performance of the crop influenced by the soil nutrient status and management factors. Chilli plant height at 30th DAP, 60th DAP and at harvest was recorded and statistically scrutinized. No significant difference was observed in plant height at 30th DAP, 60th DAP and at harvest was recorded and statistically scrutinized. No significant difference was observed in plant height at 30th DAP, 60th DAP and at harvest with respect to different treatments. At 30th days after planting, the maximum plant height of 36.71cm was recorded with application of recommended dose of fertilizers (RDF).

S. NO	Soil Properties	Value
	Physical Properties	
1	Bulk density (Mg m^{-3})	1.28
2	Particle density (Mg m^{-3})	2.63
3	Pore space (%)	51.3
	Electro-chemical and Chemical Properties	
4	pH (1:2.5)	6.88
5	Electrical Conductivity (dSm^{-1})	0.22
6	Organic Carbon (g kg^{-1})	7.9
7	Available N (kg ha^{-1})	286
8	Available P (kg ha^{-1})	41.2
9	Available K (kg ha^{-1})	358
10	Exchangeable Ca (cmol(p^{+})kg^{-1})	4.71

11	Exchangeable Mg (cmol(p^+)kg^{-1})	1.4
12	DTPA Fe (mg kg^{-1})	18.3
13	DTPA Mn (mg kg^{-1})	7.47
14	DTPA Cu (mg kg^{-1})	3.24
15	DTPA Zn (mg kg^{-1})	2.35

Table 1: Effect of yield enhancers on plant height of Chilli hyb. Arka Meghana

Treatments	Plant Height		
	30 DAP	60 DAP	At Harvest
P1 @ 150 kg/acre	31.27	86.03	96.27
Control (with RDF only)	36.71	77.51	88.73
P3 @ 3ml/lit	35.38	82.19	92.07
P4 @ 1g/lit	34.98	88.06	97.47
P2 @ 2ml/lit	33.81	84.87	96.60
P6 @ 1.5ml/lit	32.49	82.98	96.33
P5 @ 3ml/lit	34.12	83.27	95.20
P7 @ 2ml/lit	33.99	86.36	98.13
CV (%)	5.36	5.24	9.09
SE (d)	1.491	3.590	7.061
LSD at 5%	NS	NS	NS

However, applications of P4 @ 1 g per litre at 60th DAP and P7 @ 2 ml per litre at harvest recorded maximum plant height of 88.06 cm and 98.13 cm, respectively. Applications of P2 @ 2ml per litre, P6 @ 1.5 ml per litre and P1 @ 150 kg per acre also found to improve plant height at later growth stages of chilli. Application of recommended dose of fertilizers recorded minimum plant height at 60th days after planting (77.51cm) and harvest (88.73 cm) compared to other treatments. The increase in plant height is rapid during the initial stages of growth and plants attain good height within two months of transplanting (60th DAP). After this period, the growth of plants slows down as initiation of reproductive phase.

Number of branches

Number of branches has direct correlation with total fruit yield since it increases fruit bearing capacity of plant. It might be due to increase in number of leaf which has positive association with leaf area index and photosynthetic activity. Total number of primary branches at 30th DAP and 60th DAP were recorded and presented in Table 2. As evident from the data given in Table 2, there was significant difference in the number of branches with respect to different treatments.

Table 2: Effect of yield enhancers on number of branches of Chilli hyb. Arka Meghana

Treatments	Number of branches		
	30 DAP	60 DAP	At Harvest
P1 @ 150 kg/acre	7.40	9.13	9.63
Control (with RDF only)	8.33	8.60	9.10
P3 @ 3ml/lit	7.40	8.43	8.67
P4 @ 1g/lit	7.73	8.33	8.57
P2 @ 2ml/lit	7.87	8.40	10.03
P6 @ 1.5ml/lit	7.67	8.47	10.32
P5 @ 3ml/lit	8.73	8.87	9.43
P7 @ 2ml/lit	7.33	8.93	9.17
CV (%)	7.76	18.15	8.40
SE (d)	0.495	1.285	0.640
LSD at 5%	NS	NS	1.35

The highest numbers of branches at 30th DAP was recorded with foliar application of P5 @3 ml per litre. However, the trend was changed during the later stages of crop growth with foliar spraying of yield enhancers. At 60th DAP; the maximum number ++ of branches was recorded with application of P1 @ 150 kg per acre and P7@ 2 ml per litre. Application of P6 @ 1.5 ml per litre and P2 @ 2ml per litre recorded maximum number of branches during harvest stage.

Plant stem girth

Stem girth is a morphological parameter and also a major site for storage of food material from photosynthesis. Thick stem is considered to be advantageous in relation to growth and development. It is indication of extra capacity to store food material, which is useful during moisture stress situation. No significant difference was observed in plant stem girth at 30th DAP and 60th DAP and at harvest with respect to application of different yield enhancers (Table 3). However, applications of P6 @ 1.5 ml/litre and P4 @ 1 g per liter found to increase the stem girth at 60th DAP and harvest stage.

Table 3: Effect of yield enhancers on stem girth of Chilli hyb. Arka Meghana

Treatments	Number of branches		
	30 DAP	60 DAP	At Harvest
P1 @ 150 kg/acre	8.92	18.32	23.81
Control (with RDF only)	8.52	18.14	23.59
P3 @ 3ml/lit	8.25	19.27	25.05
P4 @ 1g/lit	8.51	19.39	25.21
P2 @ 2ml/lit	8.25	18.36	23.87
P6 @ 1.5ml/lit	8.40	20.56	26.73
P5 @ 3ml/lit	8.29	18.50	24.04
P7 @ 2ml/lit	8.13	18.16	23.61
CV (%)	7.91	9.40	9.40
SE (d)	0.543	1.446	1.880
LSD at 5%	NS	NS	NS

Plant and root biomass

Biomass is a plant attribute that is time consuming and difficult to measure or estimate, but easy to interpret. Biomass is defined as the sum total of life at any given time and a certain area. Biomass can be expressed as the biomass volume, wet weight biomass and dry weight biomass. According to Salisbury and Ross (1995) fresh mass is determined by harvesting the whole plants or parts of the plant and weigh them quickly before too much water evaporates from the material. Application of yield enhancers did not significantly influence plant and root fresh biomass of chilli hybrid Arka Meghana (Table 4). The highest fresh plant biomass and root biomass was recorded with application of P1 @ 150 kg per acre (601.8 g plant & 13.54 t ha) and P5 @ 3 ml per litre (63.67 g plant' & 1.43 t ha"), respectively. Highest total biomass was also recorded with soil application of P1 @ 150 kg per acre (14.94 t ha') followed by foliar application of P5 @ 3 ml per litre (14.33 t ha). Root length of the chilli plant was also varied among the treatments. The maximum root length was recorded with application of P2 @ 2 ml per litre. Application of recommended dose of chemical fertilizers was found to register highest dry plant biomass (143.04 g plant [1] & 3.22 t ha') in chilli hybrid Arka Meghana, followed by P2 @ 2ml per litre (137.73 g plant' & 3.10 t ha), P6 @ 1.5 ml per litre (137.37 g plant & 3.091 ha) and P5 @ 3 ml per litre (137.18 g plant & 3.09 t ha [1]). However, the soil application of P1 @ 150 kg per acre increased the root dry biomass (19.37 g plant & 0.44 1 ha) compared to other treatments (Table 5).

Yield and yield attributes of chilli

The results on yield parameters of chilli viz., number of fruits per plant, average fruit weight, yield per plant, yield per plot and yield per hectare as influenced by yield enhancers are presented in table 6..

Number of fruits

In the present study, fruit yield was directly associated with number of fruits per plant as number of fruits per plant increases fruit yield proportionately. Total number of harvested fruits per plant was counted and the data is presented in Table 6. The highest number of fruits (288.07) was recorded with application of P5 @ 3 ml per litre. Application of P6 @ 1.5 ml per litre and P3 @ 3 ml per litre also recorded good number of fruits (287.6 and 271.47) and found to be on par with application of P5 @ 3 ml per litre. The lowest number of green chilli fruits (235.7 was recorded with application of P7 @ 2 ml per litre

Table 4: Effect of yield enhancers on root length, fresh plant and root biomass of Chilli hyb. Arka Meghana

Treatments	Fresh plant biomass		Fresh root biomass		Total biomass (tonnes)	Root length (cm)
	Per plant (g)	Per hectare (tonnes)	Per plant (g)	Per hectare (tonnes)		
P1 @ 150 kg/acre	601.80	13.54	62.27	1.40	14.94	30.27
Control (with RDF only)	523.40	11.78	61.07	1.37	13.15	28.60
P3 @ 3ml/lit	478.93	10.78	50.27	1.13	11.91	27.33
P4 @ 1g/lit	494.60	11.13	58.13	1.31	12.44	26.40
P2 @ 2ml/lit	560.80	12.62	64.27	1.45	14.06	33.80
P6 @ 1.5ml/lit	512.92	11.54	53.80	1.21	12.75	30.87
P5 @ 3ml/lit	573.22	12.90	63.67	1.43	14.33	33.00
P7 @ 2ml/lit	541.60	12.19	59.80	1.35	13.53	31.67
CV (%)	26.33	26.33	19.88	19.88	25.07	14.48
SE (d)	115.191	2.592	9.605	0.216	2.740	3.575
LSD at 5%	NS	NS	NS	NS	NS	NS

Table 5: Effect of yield enhancers on dry plant and root biomass of Chilli hyb. Arka Meghana

Treatments	Dry plant biomass		Dry root biomass		Total dry biomass
	Per plant (g)	Per hectare (tonnes)	Per plant (g)	Per hectare (tonnes)	(tonnes)
P1 @ 150 kg/acre	136.39	3.07	19.37	0.44	3.50
Control (with RDF only)	143.04	3.22	18.52	0.42	3.64
P3 @ 3ml/lit	121.10	2.72	17.09	0.38	3.11
P4 @ 1g/lit	128.21	2.88	18.03	0.41	3.29
P2 @ 2ml/lit	137.73	3.10	18.85	0.42	3.52
P6 @ 1.5ml/lit	137.37	3.09	18.01	0.41	3.50
P5 @ 3ml/lit	137.18	3.09	17.81	0.40	3.49
P7 @ 2ml/lit	133.09	2.99	19.30	0.43	3.43
CV (%)	21.47	21.47	19.12	19.12	20.78
SE (d)	23.535	0.530	2.868	0.065	0.583
LSD at 5%	NS	NS	NS	NS	NS

Fruit weight

Average fruit weight of chilli was calculated by dividing total fruit weight per plant with total number of fruits per plant and given in table 6. Significantly highest average chilli fruit weight (6.47g) was recorded with application of P6 @ 1.5 ml per litre, followed by application of P5 @ 3 ml per litre (6.21g) and recommended dose of chemical fertilizers (6.30g). The lowest average fruit weight of 5.44g was recorded with application of P1 @ 150 kg per acre.

Fruit length and girth

Chilli fruit length and girth was measured in 100 fruits (20 fruits per plant) and average values were calculated and given in Table 6. There was no significant difference was observed with respect to fruit girth. However, significant difference was found in fruit length., Maximum average fruit girth (11.93mm). was recorded with application of P4 @ I g per litre, followed by P3 @ 3 ml per litre (11.66mm) and application of P7@ 2 ml per litre recorded the lowest chilli fruit girth (9.86mm). However, the maximum fruit length of 13.4cm was mP2 @ 2ml per litre and soil applications

Table 6: Effect of yield enhancers on yield of Chilli hyb. Arka Meghana

Treatments	Yield / Plant (kg)	Yield / Plot (kg)	Yield / ha (tonnes)
P1 @ 150 kg/acre	1.40	63.08	31.54
Control (with RDF only)	1.63	73.34	36.67
P3 @ 3ml/lit	1.60	72.09	36.05
P4 @ 1g/lit	1.42	64.01	32.00
P2 @ 2ml/lit	1.47	66.10	33.05
P6 @ 1.5ml/lit	1.87	83.96	41.98
P5 @ 3ml/lit	1.81	81.57	40.79
P7 @ 2ml/lit	1.45	65.43	32.72
CV (%)	12.07	12.07	12.07
SE (d)	0.154	6.924	3.462
LSD at 5%	0.323	14.47	7.27

CONCLUSIONS

Application of yield enhancers (P1-P7) through foliar spray found to improve the growth parameters and biomass of chilli hybrid Arka Meghana. Soil application of P1 @ 150 kg per acre also improved the growth and biomass of the plant.

Among the yield enhancers, application of P6 @ 1.5 ml per litre recorded highest per plant, per plot and per hectare yield and found to be statistically on par with application of P5 @3 ml per litre, recommended dose of fertilizers (RDF) alone and P3 @ 3 ml per litre.

Hence, the foliar spray of P6/P5/P3 along with recommended dose of N, P and K fertilizers may be recommended for realizing better yield and yield attributes in chilli.

Soil application of P1 @ 150 kg per acre along with recommended dose of fertilizers and foliar spray of any one of the yield enhancers would be helpful in realizing better growth and yield of chilli over application of P1 @ 150 kg per acre alone.

22. COMPREHENSIVE TRIAL REPORTS

Year	Crop	Institution	Control	Treatment	Yield Control	Yield with treatment	% increase in yield with treatment
1997-98	Paddy	Government of West Bengal, Office of the Agricultural Development Officer	Plot without Agromin treatment (Var.IR-64)	Plot treated with Agromin Liquid (Grade-4) (Var. IR-64)	3440 kgs/ha	4050 kgs/ha	17.73%
1992		Bangladesh Rice Research Institute	No Fertilizer (Var.BR-14)	NPK + Chelamin @500gms/ha (Var.BR-14)	3.59 MT/ha	5.13 MT/ha	42.90%
1990		Crop Research Centre Pantnagar	NPK Only	0.1% Chelamin spray	22.50 qntl/ha.	36.80 qntl/ha	63.56%
1990		Crop Research Centre, Pantnagar	NPK Only	Chelamin basal treatment 0.5 kg/ha	22.50 qntl/ha	31.20 qntl/ha	38.67%
1989-90		Director of Agriculture, Barabani, Uttar Pradesh	No Zinc Application	500 gms Chelamin per ha	29.17 qntl/ha.	35.97 qntl/ha	23.31%
1987		Indian Council of Agricultural Research, Orissa	No Zinc (Var CR 1018)	Chelamin Foliar Spray 0.2% solution at maximum tillering and panicle initiation	30.30 qntl./ha.	37.70 qntl/ha.	24.42%
1987		Indian Council of Agricultural Research, Orissa	No Zinc (Var CR 1018)	Chelamin Sol Application at 10 kgs/ha	30.30 qntl./ha.	34.20 qntl/ha	12.87%
1980-81		All India Coordinated Rice Research Project, Banaras Hindu University	Control Water Spray	Agromin @0.16% solution	34.95 qntl./ha.	40.39 qntl/ha	15.57%
1978-79		All India Coordinated Rice Research Project, Banaras Hindu University	Control Water Spray	Agromin @0.16% solution	36.99 qntl/ha	40.18 qntl/ha.	8.62%

Crop	Mustard	Soybeans	Sugarcane		Wheat		
Year	1996-97	1995-96	1995-96	1987	1987	1975-76	1992-93	1991-92
Institution	Government of West Bengal Office of the Agricultural Development Officer	Agricultural Development Officer, Joypur, Bankura	Agricultural Development Officer, Bishnupur	Central Sugarcane Research Station, Padegaon, Nira. Maharashtra	Central Sugarcane Research Station, Padegaon, Nira, Maharashtra	Khedut Sahakari Khand Udyog Mandali. Bardoli	Department Agricultural Testing and Demonstration Centre, Haldwani. Nainital	Department Agricultural Testing and Demonstration Centre, Haldwani, Nainital
Control treatment	Plot without Agromin treatment (Var. B-9)	Plot without Agromin Soil treatment (Var UPSM-19)	Plot without Agromin Soil treatment (Var UPSM-19)	Normal P&K dose and 50 kg/ha Less Nitrogen	Regular NPK fertilizer dose alone	No Agromin	No Zinc Application	No Zinc Application
Control Yield	750 kgs/ha	1800 kgs/ha.	1250 kgs/ha.	84.833 MT/ha.	105.166 MT/ha.	32,332 kgs/acre	29.88 qntl/ha.	25.00 qntl./ha.
Treatment	Pot treated with Agromin Liquid (Var B-9)	Plot treated with Agromin Soil Application (Var UPSM 19)	Plot treated with Agromin Soil Application (Var UPSM 19)	Regular NPK dose + Soil Application of Agromin @ 50 kilos/ha	Regular NPK dose+ Soil Application of Agromin @ 50 kilos/ha.	Agromin 500 gms. Foliar Spray x 4 applications per acre	Chelamin 500 gms/ha	Chelamin- Chelated Zinc-0.4 kg/ha
Yield with treatment	885 kgs/ha	1950 kgs/ha	1425 kgs/ha	119.500 MT/ha.	119.500 MT/ha.	41,569 kgs/acre	36.38 qntl/ha.	30.12 qntl/ha
% increase in yield with treatment	18.00%	8.33%	14.00%	40.86%	13.62%	28.57%	21.75%	20.48%

Crop						Tea			
Year	1990-91	1988-89	1985	1981	1981	1995-96	1995-96	1994	1994
Institution	Deparment Agricultural Testing and Demonstration Centre, Haldwani, Nainital	Director of Agriculture. Barabani, Uttar Pradesh	Office of the Agronomist, Dept of Agriculture, Rajasthan	Government Farm, Ajmer	Government Farm, Ajmer	Tea Research Association, Tocklai Expt. Station, Assam	Tea Research Association, Tocklai Expt. Station, Assam	Tea Research Association, Tocklai Expt. Station, Assam	Tea Research Association, Tocklai Expt. Station, Assam
Control	No Zinc Application	No Zinc Application	No Zinc Application	No fertilizer	NPK@ 90.30:30	400 litres water per ha, per round x 3 rounds	4 kilos Zinc Sulphate per ha per round x 3 rounds	No micronutrients	No micronutrients
Control Yield	20.71 qntl/ha	32.60 qntl/ha.	24.0 qntl/ha.	2142 kgs/ha.	4172 kgs/ha.	2910 KMTH/ha	2866 KMTH/ha.	1799 KMTH/ha.	1799 KMTH/ha.
Treatment	Chelamin-500 gms/ha	500 gms Chelamin per ha.	10 kg/ha. Chelamin Basal Application	NPK + Chelamin @10 kilos/ha	NPK + Chelamin @10 kilos/ha	240 gms. Chelamin per ha per round x 3 rounds	240 gms Chelamin per ha per round x 3 rounds	Agromin@1:500 micronutrient mixture	Chelamin @ 1 kilo per ha.
Yield with treatment	27.14 gntl/ha	44.10 qntl/ha.	47.5 qntl./ha	4773 kgs./ha.	4773 kgs./ha.	3091 KMTH/ha.	3091 KMTH/ha.	2016 KMTH/ha.	2179 KMTH/ha.
% increase in yield with treatment	31.05%	35.28%	97.92%	########	14.41%	6.22%	7.85%	12.06%	21.12%

Crop	Chillies	Tomato	Onion	Potato	Cotton	Cotton	Maize	Bajra	Brinjals
Year	1981	1981	1994-95	1994-95	1992	1980	1983	1977	1977
Institution	Govt Horticulture Experimental and Training Centre, Ranikhet, U.P	Assam Agricultural University, Dept of Horticulture	Bidhan Chandra Krishi Vishwavidyalaya Farm, Kalyani	Bidhan Chandra Krishi Vishwavidyalaya Farm, Kalyani	Ramkrishna Mission Ashram, Ranchi	Chief Agronomist, Agricultural Department, Satara, Maharashtra	Plant Protection Officer, Bhagalpur, Bihar	Agricultural Section, Aarey Milk Colony, Bombay (1 guntha = 1/40th acre)	Balrampur, Gonda, Madhya Pradesh
Control	No micronutrients	No application	No Application of Zinc	No Application of Zinc and Sulphur	No Zinc (0.03 acre plot)	Water Spray only	No micronutrients	No micronutrients	No micronutrients
Control Yield	126.1 gms./fruit	62.32 fruits/plant	148.45 qntls./ha.	148.45 qntls./ha.	550 kgs.	1893 kgs./ha.	1300 kgs./acre	15.30 qntls./6 gunthas	120 qntls./6 bighas
Treatment	Agromin spray 0.1% solution x 2 sprays at Bud break stage and 1 month after last petal fall	25 ppm Agronaa application	Chelamin @20 kgs./ha	Chelamin 20kg./ha.+ 60 kg./ha Sulphur	Chelamin 500 g/acre (0.03 acre plot)	Agromin@ 1.25 kgs/ha	Agromin sprayed twice - 27 days after sowing and 21 days after first spray	Agromin Follar Application	Agromin Foliar Application
Yield with treatment	147.0 gms./fruit	148.34 fruits/plant	172.35 qntls./ha.	220.90 qntls./ha.	600 kgs.	2475 kgs./ha.	1586 kgs./acre	21.26 qntls/6 gunthas	325 qntls/6 bighas
% increase in yield with treatment	16.57%	138.03%	16.10%	48.50%	9.10%	30.74%	22.00%	38.95%	170.83%

337

www.ingramcontent.com/pod-product-compliance
Lightning Source LLC
Chambersburg PA
CBHW020725180526
45163CB00001B/111